Germany's Urban Frontiers

Germany's Urban Frontiers

Nature and History on the Edge of the Nineteenth-Century City

Kristin Poling

University of Pittsburgh Press

"Shantytowns and Pioneers beyond the City Wall: Berlin's Urban Frontier in the Nineteenth Century," *Central European History* 47, no. 2 (2014): 245–74, reprinted with permission from Cambridge University Press, copyright © Conference Group for Central European History.

Published by the University of Pittsburgh Press, Pittsburgh, Pa., 15260

Manufactured in the United States of America
Printed on acid-free paper
10 9 8 7 6 5 4 3 2 1

Cataloging-in-Publication data is available from the Library of Congress

ISBN 13: 978-0-8229-4641-0
ISBN 10: 0-8229-4641-6

Cover photograph: Heinrich Zille. *Wiese Vor Der Turnhalle Der Gemeindeschule, Dahinter Knobelsdorffstraße / Ecke Danckelmannstraße*, 1898. Silver gelatin print photograph. Stadtmuseum Berlin.
Cover design: Joel W. Coggins

To my grandparents

Contents

Acknowledgments

In writing this book I have incurred many debts. The University of Michigan–Dearborn College of Arts, Sciences, and Letters, the Minda de Gunzburg Center for European Studies at Harvard University, and the Whiting Foundation all provided financial support for writing and research travel at different stages. I am grateful to my many academic mentors, foremost David Blackbourn, whose rigor, craftsmanship, and care provide a model of academic excellence toward which I strive. Two German Studies Association panels bookended this book's drafting and conception. In 2012 Eli Rubin organized a productive panel that helped me think about the relationship between the city, nature, and housing, which then produced a special issue of *Central European Studies* in 2014. I am thankful for his support and feedback, along with the other panel participants. In 2019 Paul Buchholz and Alexander Philipps organized a panel on the idea of the commons in German literature and history. I am grateful to them as well as the other participants for their contributions to that panel and their feedback on my contribution, which helped me think through my understanding of the wasteland commons.

Numerous librarians and archivists provided me with invaluable support and information along the way. Conducting the research that eventually contributed to this book, I worked at the Leipzig City Archive, the Lower Saxon State Archive in Oldenburg, the Paderborn City Archive, the North Rhine–Westphalia State Archive in Detmold, the Nuremberg City Library, the German National Museum, the City Archive of Mainz, the Berlin State Archive, and the Prussian State Archive in Berlin. Each provided both research support and pleasant places to work. Robert Wein at the Fotothek of the City Museum in Berlin was exceptionally helpful in identifying and providing images. Writing a book while pregnant, nursing, and fully employed means making heavy use of one's local campus library and its interlibrary loan services. I am grateful to Deirdras Jones and the rest of the Mardigian Library staff as well as the professional and kind student employees who were patient with my special requests and made my very frequent visits to the library (with or without a baby along) both pleasant and efficient.

My colleagues Anna Müller, Emily Luxon, and Josh Akers at University of Michigan—Dearborn created a supportive environment in which to take on writing as a new faculty member. Pamela Pennock and my colleagues in the History discipline provided indispensable guidance and mentorship. Georgina Hickey, Kellee Weinhold, and Denise Ho all provided writing companionship and structure. I am also grateful to my students at Harvard, the University of Rochester, and the University of Michigan—Dearborn for always reminding me of the importance of good stories and good questions. My former student Brian Kudron provided invaluable assistance with my bibliography. Without the cheerful support of my editor at the University of Pittsburgh Press, Sandy Crooms, I would likely never have found the courage to submit my manuscript.

Through our twenty years together, Mark Radosevich has given me his unfailing support, partnership, and patience, along with occasional editing and tech support. Without him this book would not have been possible. Also indispensable was the support of our family, and especially our children, Leo and Elliott. Leo deserves particular thanks for reminding me that it really is cool to write a book and for asking me if it was done yet, every day when he got home from school.

I dedicate this book to my grandparents, because I do not know their stories. I do not know what it is like to be an immigrant or a child of war. But, as I write about past visions of the future connecting German and American landscapes in a geography of journeys, I think about their lives, dreaming of open futures on another continent, taking their babies to Australia, to Canada, to Buffalo, chasing their own vision of the frontier.

Germany's Urban Frontiers

Introduction

Beyond a city's borders, real or imagined, lies its frontier: a space of anticipated growth and confrontation with the untamed. Gustav Freytag's novel *Die verlorene Handschrift* (The lost manuscript), published in 1864, opens with a portrait of an urban frontier as a distinctively German space. Freytag leads his reader through a desolate moonlit forest at the edge of an unnamed city. The wood opens, and the city appears past a meadow and a pond. Two villas, recently constructed on the edge of the forest, are the first outposts of the city. Freytag introduces one of the villa owners, standing outside his new home, delighting in his conquest of a very particular and near-to-hand wilderness. "Surrounded by light and air and free nature," the villa owner imagines himself, "the foremost pillar of civilization against the primeval forest." This begins the story of the German burgher as a pioneer frontiersman settling a new land, but this frontier is neither the Wild West of the Americas nor the "Wild East" of Poland. Here the newly conquered space at the edge of the city was also that of Germany's own storied past. As Freytag describes them, the woods, fields, and rural villages beyond the unnamed city's edge are full of the ghosts of Goethe, the Grimms' tales, and the titular lost manuscript of Tacitus. How did this city's urban edge—ringed by long-settled suburbs and villages, the stuff of centuries of German tradition—

come to seem a frontier in the wilderness and an outpost of civilization in the "primeval forest," in the tongue-in-cheek telling of one of Germany's most popular nineteenth-century novelists?[1]

The portrayal of the urban edge as a frontier is a powerful image that came, in different ways, to characterize nineteenth-century experiences of the German city and its place in a changing national and global landscape. By the end of the century, the urban edge had become a vital site for imagining the future shape of German society: a site of productivity and growth, of settlement, of confrontation with nature, of hopes and fears for the future. But the urban edge was also a site of history, tradition, and remembrance. The walls, gates, and ramparts of historic city centers embodied a vibrant past of urban self-governance, civic culture, and independence that was among the German Empire's proudest inheritances from a past shaped by regional diversity. The tensions around urbanization were often voiced as choices between city and nature, productivity and waste, ambition and nostalgia, the global and the local, choices that were confronted in the planning, design, and use of the urban edge. As examined in this book, these tensions played out in local decisions about changing urban borders as Germany's cities removed their fortifications, walls, and gates and looked for new ways to define and envision the collective horizons of urban community. Like Freytag, what Germans saw on the urban edge was both profoundly new and also tightly connected to the past.

GERMAN CITY WALLS AND THE EXPERIENCE OF HISTORY

Most German cities were once walled.[2] As both an architectural landmark and an institution, the city wall connected early modern towns large and small, giving them a common spatial structure. Medium-sized and larger cities often had elaborate fortifications built up over centuries. These included independent cities like Bremen, residences and capitals like Dresden, and military fortress cities like Ulm.[3] Smaller cities, like Mack Walker's hometowns of 750 to 10,000 inhabitants and those in Prussia and Bavaria, were more likely to have instead a simple enceinte: a wall to distinguish town from "flat land," to control the coming and going of goods and people, but not meant for military defense. In spite of these differences, through the common structure of the wall, ideas of sovereignty, territorial fixity, and exclusivity all connected towns, large cities, fortresses, and states by spatial analogy across scale. Symbolically, the wall embodied the corporative worlds of trade guilds, communal citizenship, face-to-face economic and political relationships, and a distinctively German emphasis on civic concord that all helped preserve the insularity and durability of German towns.[4]

Most of these walls disappeared over the course of the eighteenth and nineteenth centuries.[5] The nineteenth century brought visible and remarkable growth: at the beginning of the century, less than one-quarter of Germans lived in cities of five thousand inhabitants or more. By 1910 that number had risen to 60 percent. Contrary to earlier assumptions, however, it was not growth that led to the dismantlement of city walls and fortifications.[6] Instead, this physical transformation had its roots in earlier changes that themselves prepared the way for nineteenth-century expansion, including demilitarization, new commercial systems, increased population mobility, and innovations in urban administration.[7] In German urban worlds that were already changing, the Napoleonic Wars definitively unwound the protective skein in which the Holy Roman Empire had so long preserved its patchwork of weak states with a tangle of local institutional eccentricities. In his study of German defortification, Yair Mintzker argues that city walls came down in a "deluge" between 1791 and 1815, because it was then that the political structure represented by the walls fundamentally fractured, as territorial changes generalized the relationship between city and state. Cities turned away from older paternalist and particularist forms of government to new liberalized institutions. Napoleonic era reforms brought new legal structures and definitions of the city. In Prussia, Baron von Stein's city code of 1808 guaranteed cities rights of self-government as part of a slate of reforms. Cities' boundaries were defined legally rather than physically (though still in contrast to the open countryside). Overall, 350 German cities lost their military fortifications between the start of the French Revolution and the end of the Napoleonic Wars.[8]

Yet, in spite of these changes, old forms lived on. Under the German Confederation and later the German Empire, a number of federal and state fortresses were maintained or rebuilt, though they less and less resembled linear city walls. Even in cities without fortifications, defense remained fundamental to urban identity after the Napoleonic Wars. The experience of those wars ruptured old forms of patriotism and created new gendered forms of what Katherine Aaslestad terms "martial citizenship," bolstered by associational life and memorialization of war that fostered a kind of community in self-defense, even after the city's built defenses disappeared. Within this culture of martial citizenship, physical traces and remnants of walls remained significant to urban communities long after their fortifications became obsolete as defensive architecture.[9]

Once the walled city ceased to be the norm for the urban spaces of German central Europe, the traces of those old borders took on particular importance in the development of Germans' collective historical consciousness. Even (or especially) in its absence, the city wall became the structure and institution that, at least ideationally, separated a fixed city center anchored in the past from the changeable peripheries of its present and future selves.[10] The topography of vanished fortifications continued to structure how many German town dwellers experienced their urban communities into the twentieth century. Drawing on extensive oral interviews with the elderly residents of Hildesheim, a mid-sized city in Lower Saxony, Andrew Bergerson documents how significant the city's history was for the sense of community and belonging of individual residents during the period between the world wars. Bergerson recounts how his Hildesheim interviewees habitually conjured up the historical geography of their city through the convivial practice of walking "around" the city's walls, even though no circuit actually existed, and only a semicircle of parks had replaced a portion of the old fortifications land. Hildesheimers re-created the unity of the historic city through their practice of walking around walls that were no longer there. This practice connected them to a profound sense of place by making them members of a community that had learned to selectively view Hildesheim in the context of historical knowledge about its traditions and development. Taking a walk with friends or family "around the walls" was both an ordinary social ritual and a kind of historical reenactment, connecting contemporary Hildesheimers to a shared past.[11]

In fact, Germans had long been walking imagined walls as a way of connecting to their city's past and community. In 1860 a schoolteacher wrote a loving description of the parks that had replaced Bremen's city fortifications over the previous decades, highlighting how one's enjoyment of the park's natural beauty was enhanced by knowing their historical origin: "Every landscape that it offers indicates the same thing for the informed eye; the bends in the water, the changes in elevation, the systematic pattern of its layout still show hints of the former fortifications." Such descriptions highlighted spatial continuities even as they described the physical transformation of the city—leaving in place old structures that then became available for reappropriation to new symbolic forms. Here, nature, history, and urban community come together as one, but in order to see the connections one must be an "informed" viewer—part of an educated urban community of citizens who have learned about the city's history. The old gave structure to the new. By shaping, seeing, walking, and narrating the changing borders of the city, Germans fostered a sense of place by connecting to a sense of time.[12]

Long before the late nineteenth-century formation of the institutions and practices of historical preservation (the subject of valuable studies by Rudy Koshar, Joshua Hagen, and others), Germans already performed what Koshar calls "memory work" by seeing their cities in terms of historical development.[13] In *Berlin's Forgotten Future*, Matt Erlin argues that Germans recognized urban development as a historical phenomenon from the late eighteenth century onward. He outlines how everyday experiences of the city led to a more complex representation of historical time in literary texts. At issue, he writes, was "how to interpret the rapid growth of cities themselves, a quasi-natural process that seems to indicate growing prosperity even as it demonstrates the limited ability of humankind to shape its own destiny."[14] Once we think of historical awareness not only in terms of preservation but instead in terms of urban processes, we can see that it means more than remembering a vanished past; it also entails thinking about the pace and nature of change over time. In this sense, Germans learned to see evidence of processes of historical change in their urban landscapes over the course of the long nineteenth century. It was not only Erlin's literary authors who connected the experience of urban space with an understanding of historical change and narratives of progress and decline. In the following chapters, we will see how city administrators and ordinary Germans did the same—appealing to history to explain why gate taxes should be abolished or why their suburb ought to have streetlights. They argued about whether the past ought to be visible in the city, and which parts of the urban past should be legible in its physical form. By thinking in terms of processes of change, they also saw evidence of the past even where it had not been preserved.

The development of historical consciousness around urban change is especially evident in discussions about the urban border, where the pace of historical change was a subject of almost universal anxiety. The city grew too slowly, or too quickly; the wall was inevitably delayed in its removal and then dearly missed after its disappearance. The wall was so fundamentally linked to German city dwellers' sense of place and history that they used its persistence or disappearance by analogy to argue for a whole slate of other urban reforms. How city borders changed also provided an excellent way to compare cities and became a frequent venue for the kinds of interurban comparisons and competitiveness through which Germans processed the asynchronicity of urban change.[15]

Examining the transformation of urban borders in this way cuts across distinctions between place and space that have so dominated studies of city modernization. It is tempting to contrast what Mintzker calls the "vertical, corporative world of places," represented by the historical walled town, with the "horizontal, open" world of the modern city.[16] Such schematic contrasts are not true to the experiences of nineteenth-century German city dwellers, who saw the modern city as it changed with a deep awareness of the past and visualized the open spaces of the present in relation to their historical processes of development. When we exaggerate the modern transformation of "place" into "space," we risk seeing the only access to history as a kind of calcified nostalgia that does not allow for dynamic connections among past, present, and future. Instead, Germans re-created a sense of place through communal understanding of the processes of urban change. As the city changed, identifying a common horizon for the shared space of the city allowed for both continuity and flexibility amid transformation.

Both fortifications and city walls left profound legacies in spatial organization, urban institutions, modes of thought, and forms of citizenship. To many nineteenth-century Germans, becoming a modern German city meant understanding urban growth as a continual breaking down of walls and boundaries. The growth of the city seemed to repeatedly reenact both conflicts and celebrations over the walls' fall. Like other founding moments common to growth narratives, this profound communal experience was not so much over and done with as relived again and again in the development of the city and re-created in the ways Germans inscribed remembered geographies onto the city's form. In tracking these narratives, in this book I seek to map out the ideational landscape of the spaces beyond those walls—the "urban frontier"—and the uneven ways in which these spaces were marked by memories of earlier city forms. In so doing, I hope to reveal how central the particular historical geography of the city is to the understanding of progress, history, and growth that is our own legacy of nineteenth-century urbanization.[17]

METROPOLITAN IDENTITIES AND GERMAN PARTICULARITIES

By the twentieth century, one common model of the modern big city was of a space that was disconnected from place, nature, and history, representative of modernity as universal in its aspirations and a radical break with all that came before it. Illustrating this view, a character in Joseph Roth's novella *Flucht ohne Ende* (Flight without end), published in 1927, describes Berlin after the First World War as a city that "exists outside Germany, outside Europe. It is its own capital. It does not draw its supplies from the land. It obtains nothing from the earth on which it is built."[18] This view of the metropolis as a solipsistic world unto itself has captivated the modern imagination. Describing and understanding its spaces and cultures as they have seemed to dissolve old place-based connections has produced an extraordinary body of multidisciplinary literature. This research has given us a nuanced understanding of how urban spaces changed and, especially, of how individuals perceived and experienced those new spaces. Urbanites interpreted their streetscapes by and with new kinds of texts, from popular newspapers to films and photography.[19] Urbanites used these words and images both to understand their own neighborhoods and also to compare and connect their home cities to metropolitan experiences across the globe. In his study of fin-de-siècle Cracow, Nathaniel Wood argues that the train, telegraph, telephone, and newspaper connected cities in an "interurban matrix of words and images describing the modern world." Chad Bryant's research on the early railroad in the Habsburg Empire reveals how, even before 1848, trains and newspapers together connected places in a common liberal urban culture celebrating progress even as the speed of change brought anxiety about an unknown future. This "interurban matrix" brought together not just the largest metropoles and capital cities but also the smaller cities: a continental and even global matrix of metropolitan spaces that was as important to the identities and experiences of urbanites as the physical hinterlands and regional and national contexts of any individual city.[20]

Built urban landscapes throughout central Europe shared common features that helped foster this sense that cities everywhere were connected spaces. In his foundational study of fin-de-siècle Vienna, Carl Schorske argues that urban liberal professionals "reshape[d] the city in their own image," fusing utility and aesthetics in a city rearranged to place their own wealth and culture on display, while opening the streets to the free circulation of traffic. The signature project of the city's reorganization was the replacement of the massive and outmoded fortifications with the Ringstraße lined by museums, grand apartment buildings, and public institutions. A symbol of the ways in which the modern city was transformed, Schorske argues, the Ringstraße also effectively cut Vienna off from its suburbs and turned the city dweller's gaze inward, "suppress[ing] the vistas in favor of stress on the circular flow." Vienna's Ringstraße became a model for cities large and small across Europe, a new architecture of urban life that furnished the habitat of the middle-class and well-to-do city dweller. Even where such ambitious projects of urban planning were not possible, the general principles that shaped the new Vienna found influence across central Europe.[21]

On the other hand, one of the most persistent themes in German historiography has been the inextricable link between German identity and its multiplicity of distinct local places: unity expressed in regional histories and landscapes providing the vernacular, lived experiences of national identity.[22] Though more familiarly associated with the provincial landscapes of *Heimat*, this was true of urban spaces as well.[23] A growing literature on cities throughout the Habsburg and German Empires details the ways in which the features of the spatial and aesthetic model of the modern city interacted with local cultures and histories to produce a rich variety of "provincial" modernities from Hamburg to Lemberg.[24] Although the focus on urban history has in some ways helped liberate nineteenth-century urban histories from too exclusive a focus on national identities and revealed the many "other modernities" of central Europe's diverse urban places, in its emphasis on liberal culture and the "interurban matrix" this focus has also, like Vienna's Ringstraße, turned its back on the most immediate contexts of urban spaces.[25] In the modern bourgeois city, nature, locality, and history are represented as decontextualized symbols, not as connected experiences. Focusing on the urban border, as in this book, also reveals that "interurban matrix" on its ragged edges. Here, on the urban frontier, the metropolitan German encountered other forms of spatial organization in terms of class, planning, and the use of natural resources.[26]

THE STATE OF NATURE ON THE URBAN BORDER

Joseph Roth's peripatetic narrator imagined the big city "obtain[ing] nothing from the earth on which it is built," but the programmatic statement of urban environmental history has been William Cronon's assertion that "Before the city, there was the land." In other words, one cannot understand the city without understanding the features of the land on which it is built. The modern city poses a particular challenge to incorporating environment into its cultural and material histories. City dwellers themselves were often intent on erasing—or never seeing—the connections between city and land. Research on nineteenth-century German urban expansions, on the other hand, has focused on the technical and administrative perspectives of city planning and infrastructure. A growing city confronted nature on its borders in new and challenging ways involving resource use, leisure, and the physical conditions of settlement. Along the margins of the city, projects for urbanized green came into contact and often conflict with ideas of nonurbanized nature beyond the urban edge. Was nature to be part of the growing city, or its horizon?[27]

The breaching of city walls from the late eighteenth through the nineteenth century brought with it the integration of new kinds of green space into the city, including promenades, gardens, parks, forests, and decorative water and trees. In cities like Frankfurt, Münster, Leipzig, and Mainz, belts of promenades and parkland replaced the wall circuits.[28] The removal of city walls and opening of city gates responded to a perceived need for those most fundamental and pervasive elements of the natural world—the light and air (*Licht und Luft*) that dominated nineteenth-century German discussions of urban reform and expansion. Barry Jackisch shows us how available green space became "part of a larger solution to growing concerns about urban hygiene," echoing earlier campaigns for urban beautification.[29] Constructing access to nature became one of the technical tools of urban planning, beginning with the planning of gardens in former fortifications lands at the beginning of the century and developing into schemes for urban health hygiene at its end.

The first proposals for systematically providing access to nature in the city, whether in the form of allotment gardening, urban forests, or parks, recalled traditional uses of the urban hinterland. The aim was to reconstruct an older urban–peripheral relationship perceived to have been lost in the process of too rapid urban growth. The best-known of these is Countess Adelheid von Dohna-Poninski's proposal for a green belt of park space separating the city core from workers' colonies on the other side, giving everyone easy access to green. As a number of observers noted at the time, the green belt project naturalized earlier urban planning patterns. They suggested that cities without a park belt from defortification needed to right this wrong by intervening in the city's free development. Other urban planners and reformers resisted this suggestion—precisely because it recalled the ways fortifications hemmed cities in.[30]

As city administrators and reformers found new ways of integrating green space into urban planning practices, contemporaries increasingly defined the modern city as being antithetical to all that was natural and wild. This changed definition brought with it a compulsion to identify a landscape beyond the city that could provide the kind of uncontrolled nature the city could not. Hence, in Berlin, urban peripheral settlers appeared as frontiersmen returning to a state of nature, and the Grunewald hunting preserve was transformed into a primeval forest. In his rambles through Berlin's hinterland, with his notes published in 1861, Theodor Fontane invented the lost pre-reclamation Oderbruch as a wild space. As planning practices became more universal, Germans began imagining spaces beyond the reach of planners as spaces of not only disorder but also (lost) freedom.[31]

In many German cities, urbanites understood the premodern walls and boundaries themselves to reflect the physical preconditions for a given city's economy and community—topography, access to water and resources, etc.—

that allowed urban flourishing in a particular spot. Walls and physical borders came to represent the historical relationship between urban and natural landscapes in the symbolic geography of the city. Changing city borders—in an era of urban expansion and modernization—also challenged historically rooted understandings of how urban communities shaped and depended on their physical regions. In this way, in the nineteenth century urban border systems, either built or unbuilt, became elements of landscape in the modern sense: as representations of spatial relationships they "act[ed] to 'naturalize' what [was] deeply cultural."[32]

METHOD AND ORGANIZATION IN THIS BOOK

Analysis of a detailed set of case studies from different geographical regions distinguishes this book from other works on the development of German urban spaces, which have most often focused on single cases or particular regions or have generalized across the German urban landscape. The wide range of sources for this project—from archival research in municipal and state archives to extensive periodical research and the published writings of German social thinkers, urban reformers, and planners—allows us to investigate the process of urban expansion both as a local story embedded in the politics and environments of particular cities and as an emerging discussion among urban experts in the German Empire and across the globe. Through the five chapters, an interconnected set of case studies provides a nuanced view of the relationship between local circumstances and shared notions of urban space and development. Cities such as Oldenburg, which have been the subject of less sustained study, highlight the regional diversity of German urbanism and the environmental aspects of urban growth and add nuance to the standard narrative of German urban development, which is dominated by a few urban centers. The case study method adopted here is an analytical choice based on the belief that environmental and historical cultural dynamics of urban development are best revealed through the specific topographies of individual cities.

Chapters 1 and 2 take Leipzig, the Saxon market and university city, and Oldenburg, the northwestern ducal residence, to lay out the political, social, and environmental stakes of urban border changes over the long nineteenth century, as these cities were embedded in distinctive local landscapes. Urban borders regulated the movements of people and goods and the visibility of social and political hierarchies on the one hand; they also arose from a constantly renegotiated balance of power between city and hinterland, nature and built environment. These first two chapters demonstrate that opening urban borders was neither as simple nor as quick as leveling walls or unlocking city gates. Chapter 3 examines the persistence of the city wall through the middle decades of the nineteenth century, both as a physical feature of the urban landscape and as a potent symbol for the urban past. City and state negotiated a shifting balance of power and responsibility in the struggle over urban excise taxes and over who was to care for Prussia's many crumbling walls during their growing obsolescence from the 1830s through the 1880s. In both the capital city of Berlin and the small Catholic town of Paderborn in Westphalia, contemporaries projected anxieties about delayed modernization on the persistence of walls that came down "too late," but they deployed that narrative of delay in the service of different spatial stories—one of centrality and the other of marginality.

Chapters 4 and 5 examine how the urban border became a space for negotiating the relationship between the city's present form and an anticipated future and a historical past in the increasingly connected urban landscapes of the German Empire. Beyond Berlin's tax wall, which was removed in the 1860s, there emerged new visions of the urban future, horizons of future development that themselves became the (often imagined) visual border of the city. As Berlin took its place as a united Germany's new and not always very popular capital, these new images of the urban edge and urban growth carried national resonance. On the other hand, emerging movements for the preservation of built and natural landscapes invested historical border systems with new significance. City walls made for particularly vexed objects of historical preservation campaigns because they invoked competing narratives of continuity and rupture with the past that were themselves each a key to modern German and metropolitan identities. This tension played out in the national uproar over the planned destruction of Nuremberg's city walls in the 1870s and 1880s. Communal visions of the urban past and urban future competed on city borders.

In each chapter, a local case study is used to open up a thematic discussion that intersects with the other chapters. This method reflects the process of urban development itself as individual cities were increasingly netted together by national and global ideas of what it meant to be a modern urban space. In these new conversations and exchanges of urban expertise, local particularities of landscape and history that had once defined cities' urban identities became either challenges to be confronted by urban planners and administrators or treasured inheritances, the value of which were determined by translocal marketplaces of travel, historical preservation, and tourism.[33]

If "history essentially begins with differentiation between the present and the past," as Michel de Certeau has written, nineteenth-century urbanites often looked to the moment when the city burst through the shackles of its walls to mark this moment of differentiation—the origin story of the city as a modern community. The relation of the modern city to the walled city of the past had (and has) particular trenchancy, because the spatial development of the city models the fundamental structure of historical narration itself. The past is a closed space of experience, and the future opens up a horizon of expectation. As Arthur Danto has described it, the practice of history is that of accounting for a closed past at the same time maintaining the possibility of an open future. This could equally well describe the story nineteenth-century Germans came to tell, with pride and sometimes anxiety, about their cities. In understanding and narrating the spatial development of the city, nineteenth-century observers and planners sought to find continuities between the premodern and the modern city without foreclosing the possibility of unanticipated future development. This is perhaps also the most important meaning of the frontier: the boundary between what has been done in the past and what might be done in the future.[34]

Telling the story of the city entailed describing the development of its frontiers but also the formation of bridges to the past. As in Freytag's novel, the growing city had to conquer not just space but the ghosts of German pasts in both literature and memory. The very struggle to manage and interpret the uneven and incongruent ways in which cities grew and overcame their historical boundaries reveals the developing importance of that moment of differentiation as a communal myth that persisted through the nineteenth century and beyond.

1 Picturing the City

Urban Panoramas on the Leipzig Ring

In the spring of 2015 the Leipzig City Library put on an exhibit entitled "Panorama of the Leipzig Ring, 1850/2015." The exhibit and accompanying book contained a series of photographs of twenty-first-century Leipzig taken along its ring road, stitched into panoramic images alongside drawings of the inner city's edge from the middle of the nineteenth century (see figure 1.1). On one hand, glassy facades and wide roads contrast with the lawns, linden trees, and lively strollers in the earlier images, but on the other hand, also impressive is the degree of stability between these views of the city, separated by over 150 years. They share both the broad outline of the city's built form and a number of key urban institutions—from the University of Leipzig to Thomas Church, where Johann Sebastian Bach once toiled. As the exhibit catalogue puts it: "Along the promenade ring Leipzig traditionally puts its best face forward."[1]

The Leipzig Ring is a space originally created by Leipzig's process of defortification, a wide space left behind when the city's walls, moats, and ramparts were removed and gradually replaced by parks, roads, market squares, and new urban institutions. A few traces of the former defensive infrastructure remain—in a pond and park where once there was a moat, in the distinctive outline of one of the former bastions (now home to a cultural center and

1.1 Jörg Dietrich's street-line panoramic views of the Leipzig Ring were displayed alongside drawings from a historical panorama of the promenades in an exhibit at the Leipzig City Library in 2015. Dietrich's photograph above shows the stretch of the Ring by the new city hall. Reprinted with permission of Jörg Dietrich. Below is the same stretch in the 1850 panorama, with the Pleißenburg Fortress on the left. Reprinted as *Der Leipziger Promenadenring zur Biedermeierzeit* in 1989. Original in possession of the Stadtgeschichtliche Museum Leipzig.

café), and in a single tower preserved from the city's fortress. Although the Leipzig Ring today is a massive road that circles the city's historic urban core with spacious medians, crisscrossing streetcar lines, and branching entrances to underground parking, it remains in some ways an outer face for Leipzig, a place where the city presents itself to the world, putting on display both its contemporary vibrancy and its connections to the past.

Just as the 2015 exhibit displayed Leipzig's change over the previous century and a half, the images of the Leipzig ring from 1850 would on their own have evoked a historical process of transformation familiar to the city's residents of that time. The demilitarization of the city's edge, the dismantlement of the fortifications, initiated in 1770, and the construction of promenades and English-style gardens in their place, all allowed the creation of the Leipzig Ring as a space for picturesque walks and panoramic views of civic life.[2] The ring followed the circuit of shaded promenades along its former moats, revealing the city's university and churches, as it still does today, along with since vanished urban institutions such as the city's Bürgerschule on the Moritz Bastion. As in the 2015 photographs, the 1850 panorama contains a patchwork of old and new: stretches of ramshackle city wall remain alongside the stylish promenades, and less orderly spaces are conveniently hidden by the topography of the mostly drained but not yet entirely leveled moats. The process of urban transformation is on display.

Beyond highlighting the ongoing importance of the Ring as Leipzig's outer face, the exhibit suggests a way of viewing images of the city that a historian of photography, Catherine Clark, calls "repicturing"—that is, seeing the contemporary urban landscape in the context of images of the city's past.[3] The glass and stone surfaces of the Leipzig Ring today are overlaid by images of the promenades of the city's past, investing the contemporary landscape with new significance in its connection to the past. This way of viewing the city has particular relevance for our understanding of the relationship between cities and their borders over the course of the long nineteenth century, during which time Germans learned to view the city edge in the context of a history of urban development: seeing and interpreting the city with their knowledge of how it looked in the past.

From the beginning of urban defortification, Germans worried about the ways in which the past would—or would not—be visible in the changing city. They saw new and changing spaces overlaid by familiar images of the city's past forms. Clark traces the emergence of repicturing in the specific context of the popular reception of and participation in photography of Paris, but the concept has broader usefulness in understanding the ways in which nineteenth-century urbanites perceived and narrated urban change. Decades before urban preservationists encouraged Germans to "revisualize" ordinary buildings and townscapes as "documents in stone," as Rudy Koshar puts it, Leipzigers were arguing about physical changes in their city, understanding those alterations in relation to a historical process of change and debating whether they wished to see that urban past preserved or discarded—particularly when it came to the city's borders.[4] Contemplating the city's relationship to the past was part of the German experience of urban modernity, not just with the rise of popular history at the end of the century but from the beginning of the modern urban transformation.

It is no accident that the format chosen for this repicturing of the city in the space of the Leipzig Ring was the panorama, a technology of representation with close connections to the transformation of the modern city. An invention of the eighteenth century, the panorama instructed viewers on how to look for the whole and construct a totality out of complexity, whether in the built environment, in the natural world, or as a historical scene. A popular form of urban spectatorship in the nineteenth century, the panorama supplied a unified visual identity that the physical city threatened to sacrifice to growth and the loss of clear borders. The urban panorama itself served as a kind of urban edge—a surface of contact between the city and the outer world, structured by sight lines and horizons in what Renzo Dubbini has termed the "geography of the gaze."[5] Dubbini and others have argued the importance of this visual geography in maps and paintings, and a close archival history of the city of Leipzig reveals the ways in which a related logic of sight lines and horizons shaped how ordinary city dwellers understood and made decisions about their changing urban built environments.

Like panoramas, city borders functioned as ordering structures through which nineteenth-century urbanites viewed and approached the city.[6] In this way, the space of the Leipzig Ring functioned much as the panorama did as a form of visual representation. Visual representations of the city—the urban panorama, the *venduta*, the urban tour—guide their viewers in how to see the city, and the parks, promenades, gates, and peripheral spaces of the city edge likewise organized the view of the city that was available to insiders and outsiders alike. But unlike a painting, usually the vision of a single artist, the spaces of the urban edge were the product of conflict and disagreement among people of competing social groups and interests who had different understandings of the city's status as a space for communal identity and action. Leipzig's walls and fortifications did not stand or fall by a single decision of either the Saxon state or the city government: they were hidden, obscured, selectively leveled and rebuilt over a century: the past, present, and future of the city knitted together in the geography of a stroll, in the framed view of strategically placed linden trees, in the fragments of wall left standing along the promenades, and in selectively chosen vistas. Each small decision was part of the process of making and unmaking the urban edge as a border and an outward face of the city—a process that, as revealed in the 2015 exhibit, continues in new forms into the twenty-first century.

A WALK INTO THE EIGHTEENTH-CENTURY CITY

Leipzig is the site of the most famous literary scene in all of German literature involving the exiting and entering of a city. The second scene of Goethe's *Faust*, written in the decade of Leipzig's defortification, opens with a descriptive panorama of town life as seen just outside the city gates. On a spring evening, the town's occupants emerge from the city to enjoy the good weather. Leaving the town is synonymous with leaving the world of work and order. Young apprentices head for the taverns. Burghers discuss politics. Farmers drink and celebrate. Girls stroll about to see and be seen. Released from the limitations of town life, the atmosphere outside the city gates is one of celebration and cautious abandon, distinguished by a disorderly mix, a "bunter Gewimmel," of classes, occupations, ages, and genders.[7] Faust oscillates between claustrophobia within the city walls and a dizzying agoraphobia brought on by the open fields outside of the city gates. His recurring fantasy is to see the earth from a bird's-eye view, and so to encompass with his gaze both the interior of the city and the open land beyond its walls. In contrast, his love interest, the virtuous Gretchen, explains to Faust that what every good burgher dreams of is a small garden along the city wall, not bird's wings or permanent escape from the city. Goethe himself participated in the defortification of the city of Jena, in Thuringia, not as a grand Enlightenment project but, rather, as a gradual construction of footbridges and garden plots in the shadow of slowly crumbling stone walls.[8] What most profoundly connects his literary vision of Leipzig with its real-world landscape is the importance of sight: a longing to observe the city and to free one's sight from the limiting horizon of a wall.

Although holding such an important place in Germany's literary landscape, the environment around the Saxon city of Leipzig is less monumental. The city sits in the lowlands where the Pleiße and Parthe Rivers join the course of the Elster. To the west is a marshy landscape of riparian woodlands known as the Leipziger Auwald. The surrounding land is smooth and fertile, dotted by lakes to the south and northwest, either peaceful or monotonous, depending on one's point of view. One city guide published around the turn of the eighteenth century described the landscape as pleasant but regrettably flat.[9] In this low landscape of lakes, small villages, and forests, what face did the city of Leipzig present to the outside? This was a question Leipzigers asked themselves frequently as changes in the city's character, economy, and physical landscape transformed its relationship to the immediate hinterland, Saxony, and the wider world. In this process Leipzigers considered the view of the city both from its edge and as approached by travelers from afar.

Leipzig was the "second city" of Saxony, a midsized state and a rival of its larger and growing neighbor Prussia. Politics in the state tended to the conservative. In the nineteenth century, that Saxon tradition of conservatism was challenged by rapid population growth and industrialization, which would also make some of Saxony's cities home to a large, organized, industrial working class. Nonetheless, as Abigail Green has shown, this conservatism did not hinder Saxony from adapting creatively to the challenges of modernization.[10] The capital of Saxony and residence of the Saxon royal House of Wettin had long been the much more stylish city of Dresden, about one hundred kilometers east of Leipzig.

An old saying has it that "In Chemnitz money is earned, in Leipzig it is increased, and in Dresden it is spent." Saxony's urban landscape thus conveniently divided into workday industrial center, marketplace, and glamorous capital. This is a useful reminder that cities are best viewed within regional networks, but the clean division of urban roles does not help us understand the place of Leipzig in Saxony throughout its history.[11] Leipzig was not only a market city but was also home to a significant university, a center of the book trade, and later an important industrial center. As a university town and a market center, Leipzig seemed virtually independent of the state and free from the pretensions of court or nobility. Historian Robert Beachy has observed that Leipzig's city government operated more like an independent Hanseatic city than a territorial Landstadt.[12] Leipzig's university, book fairs, and publishing industry gave it a central place in German intellectual life. Its trade fairs and prominent mercantile elite established its significance for continental trade. In competition with citizens in places such as Frankfurt am Main, Leipzigers fancied their city to be Germany's commercial heart and a capital of culture, trade, and fashion.[13] Leipzig's scholarly and business classes comprised its own kind of local royalty. Far-flung commercial networks and the global merchandise these networks funneled into the city's yearly markets lent Leipzig a cosmopolitan character that transcended its Saxonness. In his autobiography, Goethe wrote that when he arrived in Leipzig as a student in the 1760s, he found a comforting familiarity in the trade fairs, with their "well-known wares and traders; only in other places and different order," though the goods and traders themselves came from exotic realms to the east and the south. Leipzig's cosmopolitan character, early recognized as a distinguishing trait, also made the market city indispensable to Saxony's economic interests.[14] These relationships, both economic and political, became inscribed in the city's complex and multipart urban border system, marking them in the landscape of the city and in the experiences of both city dwellers and visitors.

A traveler approaching the city in the early eighteenth century would have come first upon a patrol at the edge of the city's imperial market privilege, marked by a fence. What he saw on the other side of this simple outer border would not have looked much like a city. The areas of the city outside the moats and ramparts of the urban core had a largely rural character. Here, those traveling by either foot or wheel would be asked to declare all goods subject to excise taxes. Failure to do so could result in heavy fines, but visitors who did not plan to stay long had the option of leaving their luggage behind at one of these outer gates for safekeeping. The gate attendant (*Torschreiber*) would ask anyone entering the city for their name, occupation, origin, and destination.

The names of all visiting foreigners were then compiled in a daily list, which was conveyed to guest and coffee houses and, in later years, published daily in the newspaper.[15]

The gate attendants would not let just anyone past their gatehouses and into the city. Leipzig had the reputation of suffering under a severe "scourge of beggars" (*Bettelbedrängnis*) in the eighteenth century, a problem arising in part from its yearly markets, which offered the prospect of temporary employment for many impoverished Leipzigers and visitors. The city's outer border played a key role for the enforcement of Leipzig's laws regulating beggars. Gate attendants were instructed to keep out foreign beggars, who had no right to alms in the city.[16] Among those the gatekeepers singled out for particular attention were "suspicious pregnant women" suspected of seeking entry in order to baptize their illegitimate children in city churches. Jewish traders and fairgoers were required to obtain a special pass to enter the city from 1668 to 1764. Contemporary observers blamed the presence of beggars within the city on the gatekeepers' ineffectiveness. In 1748 and again in 1772, the city council threatened to punish gatekeepers who did not sufficiently examine "alien and sick persons" entering the city.[17]

In the eighteenth century the land between this outer border and the inner city was neither legally nor popularly considered part of the city proper, though Leipzig's suburbs had a long history of settlement with well-established social and economic ties to the city. A city guide from 1792 suggests that Leipzig's suburbs were to the city proper as the introduction was to a book: necessary, but ancillary.[18] A traveler approaching the city from the south would find that introduction to the city full of guesthouses, private pleasure gardens, and splendid villas. A traveler approaching along Gerber Street from the north, and crossing the Parthe, would find settlement sparser, younger, and interspersed with tanneries and tobacco production. Between the Pleisse and the Elster was the Rosenthal forest, purchased by the city from the Saxon prince in 1707. In the west along the Pleisse, our traveler would see mills and textile production, a mixed terrain of agriculture and small-scale production, characteristic of pre- and protoindustrial suburban landscapes.

Once our traveler approaching the city had made his way through the outer regions, if he wished to enter the city proper behind the moats and ramshackle fortifications, he would have to proceed through one of the city's gates. There were four main gates—Rannstadt Gate to the northwest, Halle Gate to the north, Grimma Gate with its distinctive rounded tower to the east, and Peter's Gate to the south—as well as several smaller gates open only to those traveling on foot. Peter's Gate was Leipzig's largest and grandest entrance, constructed in baroque style in honor of Elector Friedrich August I in 1723. Outside of each gate a short bridge spanned the city's moats. More mud than flowing water filled their banks except to the west where water entering from the Pleiße River deepened and quickened their flow. Here, Leipzigers sometimes fished for carp.[19] The inner gates were guarded by the city garrison, controlled by the city council. These gates would have been buzzing with traffic around the time of Leipzig's yearly market fairs, as merchants poured into the city to sell their wares.

The inner city was quite small. A stroll around its surrounding moats along a narrow linden-lined promenade could be comfortably completed in about three-quarters of an hour, and both Leipzigers and visitors often did just that. The gradual greening of the city's fortified edge began in the early years of the eighteenth century, when the first trees were planted along the ramparts.[20] The tree-lined avenue between the small Barfuß Gate and Thomas Gate quickly became a popular destination for daily walks. The new promenade was known as the "Aunties' Place" (Muhmenplatz) for the number of nurses who brought their charges there for fresh air and exercise "around the gates," but it quickly became a place where the city's distinguished residents went to see and be seen.[21] A number of beautification projects followed, including the introduction of swans to the moats outside Rannstadt Gate in 1732, and the promenades were extended around the remainder of the city in the following decades (see figure 1.2).[22]

Although a point of civic pride, the promenades remained a space of anxiety concerning borders and their transgression. Critics of Leipzig's decadent fashion-obsessed society asked whether the promenades displayed an immoderate desire for the appearance of luxury, even in times of poverty. Just as burghers tightened their belts during the week so that they could display themselves in all their finery on a Sunday stroll, the city hid its inner dreariness behind a luxurious cloak of greenery. The promenades were a place of social and moral transgression; their seediness reflected the failure of the aesthetic project. Satirist and scholar Detlev Prasch lampooned Leipzig's proud "Allee" as a dusty path, which, though decorated with lindens, actually gave one a view only of walls and houses. The foul smell arising from the sludge in the moats marred the pleasure of a stroll outside Leipzig's walls, and the mix of social classes made them a threat to the city's morality.[23]

The promenades remained in important ways conceptually outside the city. The laying of the promenades did not itself open the city or erase the border between the inner city and its suburbs. Behind the trees and gardens, the civil functions of the urban border remained unchanged. Gate taxes and

1.2 Leipzig's eighteenth-century promenades were a site of recreation and conviviality, often depicted in illustrations and city guidebooks. Here you can see how the space was both public and secluded, cut off from its surroundings, with the city wall on the right. Johann August Roßmaeßler, *Leipzig, Die Promenade von der Barfüßer Pforte bis zum Thomas Thor in Leipzig*, 1775, woodcut, Sächsische Landesbibliothek— Staats- und Universitätsbibliothek Dresden/Deutsche Fotothek. http:// deutschefotothek.de/documents /obj/30117319/df_hauptkatalog_0113118/.

walls remained. The gates were closely policed. Only the military significance of the borders disappeared in this "civil conversion process" of the former fortifications land into space for leisure and use. The place of the promenades inside or outside the city remained ambiguous even after the land beyond had been fully incorporated.[24] Nonetheless, in coming decades, Leipzigers would appeal to the promenades' appearance of openness to demonstrate Leipzig's character as an open place and to support demands for ever greater freedom of movement in and out of the city.

Leipzig's promenades were among the earliest urban beautification projects of their kind, and they quickly became one of the city's most distinguishing characteristics and a point of civic pride. Behind the trees, Leipzig's eighteenth-century fortifications would have presented a rather forlorn prospect to the approaching traveler. The walls had probably first been constructed by the tenth century, before Leipzig received its city charter in 1165. The city's walls were destroyed, then rebuilt and extended in the sixteenth and seventeenth centuries; earthen ramparts were then added after the Thirty Years' War. Of the city's early fortifications, only the Pleißenburg Fortress by Peter's Gate survived. Its bulky form dominated the city's profile, and its round central tower remained the first structure visible from afar. These extensive fortifications were once among Saxony's most impressive but later became difficult and expensive to maintain. By 1770, after the Seven Years' War, Leipzig's fortifications were "so decayed that one could no longer discern their former disposition," the outer bastions mere "unformed heaps of earth."[25]

Leipzig's eighteenth-century city border was not just a wall easily torn down. The urban border was instead a complex transitional zone composed of many different spaces and institutions that regulated the relationship between the city and the wider world. Deconstructing that border would a complex process. Even in *Faust*, the zone in-between is as important as the contrast between inside and outside. Life is spent neither entirely within the walls nor entirely without them but also within Gretchen's small piece of green, her garden on the city's edge. The outer tollbooths, the moats, the ramparts, the city gates, the inner-city wall, and even the suburbs with their gardens and forests, all formed the organizing structures through which Leipzig's residents and visitors both approached and viewed the city. Dismantling the city's fortification made visible just how the city edge worked to structure the ways in which both Leipzigers and outsiders saw the city.

OPENING THE MARKETPLACE, DESIGNING A GARDEN

From the first removal of Leipzig's ramparts in the eighteenth century to the opening of its gates and the leveling of its moats in the nineteenth century, the changing configuration of Leipzig's borders reflected concerns about how the city would be perceived, balancing the interests of the city as a marketplace and as a civic community. After the Seven Years' War, the Saxon Restoration Commission hoped that Leipzig's defortification would "restore the trust of foreign and domestic merchants."[26] When the city council petitioned Saxon Elector Friedrich Christian for permission to remove the fortifications in 1763, transforming Leipzig into a completely open place (*ein gantz offenen Ort*), it referred to Leipzig not as a city but as a center of commerce (*Handelsplatz*), basing its claim to openness on principles of trade and not on principles of urban independence, character, defense, or governance. Although changed military circumstances allowed the removal of Leipzig's fortifications, the primary concern of both state and city in managing that removal was the relationship between the city's borders and its character as a Handelsplatz and an open trade city. Both Leipzig's city council and the Saxon Restoration Commission expressed concern that the fortifications betrayed Leipzig's true identity as an open space, driving outsiders away instead of welcoming them in, as a free marketplace ought to do. As one later commentator would colorfully put it, oddly inverting the spatiality of the closed city, "a fortress in a marketplace is the wolf in a sheep's pen."[27]

After the Saxon state gave Leipzig permission to remove the remaining fortifications, competing plans for the fortifications land—including rejected plans for housing and an oilcloth factory and a successful proposal for a theater—anticipated later debates over whether the Leipzig Ring should be a space of industry, leisure, or commerce. Under the oversight of Johann Carl Friedrich Dauthe, director of Leipzig's drawing academy, and Mayor Karl Wilhelm Müller, the bulk of the demolition work was carried out between 1784 and 1794. The stone and earth removed were used to fill in some sections of the moats, leveling the ground. Other stretches, like those around Rannstadt, Pleißenburg, and Peter's Gate remained for decades as either wet moats or dry depressions. In the following years, Dauthe designed English-style gardens and extensions of the promenades, which became among Leipzig's most admired and distinctive urban spaces.[28]

As work on Leipzig's English-style gardens began, a well-known German proponent of English garden planning, Kiel Professor Christian C. L. Hirschfeld, published a popular treatise on the subject. As Hirschfeld described it, the delicate balance between enclosure and openness expressed in the ideal garden bore striking similarity to the conundrum faced by the trade city in the construction of its new, more welcoming visage. Hirschfeld wrote that "broad vistas always engage the imagination most pleasantly," because of the "soul's inherent disposition toward expansion," but cautioned "you should no more want extended vistas everywhere in a garden than in nature. Unbroken prospects on all sides are distracting and ultimately tiring. . . . Nature closes her landscape views with swells and woods; the garden artist can use buildings as well."[29]The English garden planning that was in fashion in the late eighteenth and early nineteenth centuries negotiated between the border and the vista, much as did contemporary city planning.

The gardens planned along Leipzig's edge as laid out by Dauthe served to soften and obscure from view the harshness of the walls, without eliminating the city border. Borders, Hirschfeld wrote, "should not have noticeable precise edges; they are more agreeable if they merge gradually into the open landscape, without a clear boundary marked by a wall or ditch. In this way a garden looks not only more natural but also more spacious." The goal of the English garden as delineated by Hirschfeld was much the same as that of the city, which was to be closed but never to affront or restrain the eye with the too obvious appearance of a border.[30] Richard Sennett has taken the Enlightenment era fondness for the invisible border (epitomized by the haw-haw, a sunk fence or a ditch with a fence at the bottom meant to keep cows in the appropriate pastures, while hiding the fence from view) to indicate the incapacity to acknowledge and incorporate difference within community. Leipzig's promenades were an early and prominent example of a wave of such projects in cities across Germany that were carried out under the influence of Hirschfeld's philosophy, around 1800 and at the end of the era of absolutist, state-directed city planning.[31] In the layout of Leipzig's promenades, the hide-and-seek game of planning sight lines and vistas that was characteristic of English garden planning took on a second layer of meaning in the conscious manipulation of sight lines at the inner city edge.

Although intended to make the city more welcoming to visitors, the leveling of Leipzig's old fortifications also brought concerns that the wrong kind of person would now be able to gain easy entry into the city. In 1770 when first considering the details of the plan, city administrators had advised that the city still be fully closed in the evenings even after the demolition was completed in order to protect against the "creeping in" (*Einschleichung*) of persons who would "make ill use of a complete opening of the city," disturbing the peace and security of residents and smuggling in goods to avoid urban excise taxes. Similarly, in the proposal for the demolition project, Dauthe cautioned that once ramparts had been replaced with gardens, the city wall must again "be built up against the entry of dissolute rabble." Even where the moats were only partially dried, between Rannstadt and Barfuß Gates, nearby occupants complained that this gave the "lurking rabble" outside the gates "a convenient opportunity for nighttime burglary" and requested that this "imminent danger" be averted either through a deepening of the moat or an extension of the city wall.[32]

These concerns notwithstanding, the removal of Leipzig's ramparts did little to change the relationship between the city's inner core and its outer suburbs. Nor had the old moats been erased from the city landscape. Substantial portions were left unfilled, notably to the south of the city around the Pleißenburg Castle, Peter's Gate, and the Moritz bastion. By 1799, the only wet moats that remained were between Rannstadt and Barfuß Gates, but even where substantial amounts of earth and stone had been dumped into the swampy moats, the ground had not been leveled.[33]

A significant shift in the function of the border came in in 1824 when, with permission from the Saxon king, the city eliminated the nighttime closure of its gates. Although Leipzigers, and particularly students, had been protesting against the gate tax and nightly closure of the city gates for several years, the decision came instead in response to something of a public relations crisis. Leipzig's burgher committee was especially irked by a satirical commentary published in 1823 in Dresden. The article was styled as an overly enthusiastic account of Leipzig's many cultural and economic advantages, in which two visitors ironically celebrate the city's charming backwardness. They report with mock relief that in Leipzig—unlike so many other cities—there was no danger of the gates remaining open twenty-four hours a day. "The nightly separation of our good city from her flourishing suburbs" was instead preserved as "a fine old custom." They are relieved that, "To date, the good old days have been preserved and we have besides the pleasant hope, that, according to the laws of stability, suburban dwellers and foreigners will continue to be permitted the privilege of paying their little bit for the lighting of the city, in that they, as once the Jews, must pay a head tax to pass after a certain hour through the city gates, now decorated in the newest Boethian style, artfully and tastefully, and at great expense." In the article, the fictional brother travelers make light of Leipzig's cosmopolitan ambitions and link the nightly closure of the gates to Leipzig's old-fashioned and oligarchic style of government.[34]

Most of the city wall and inner-city gates came down in the 1820s and 1830s, though significant portions remained in the south and west and some pieces were incorporated into adjacent structures. There were conspicuous exceptions to the removal of the remnant of the wall as a symbol of an open marketplace. The city's Jewish market, which Robert Blum described in 1834 as an "Eldorado of planks" (*Breter-Eldorado*) on the moats outside the Halle Gate, was moved to a new location in 1838. When the market moved, the city council decided to rebuild the city wall along its edge, to be covered with ivy, reestablishing a visual border. The last gate to come down was the grand Peter's Gate, which was not taken down until 1859–1860.[35] Many of the remaining depressions and elevations where moats and ramparts had once been likewise were not leveled until debates over the leveling (*Ausfüllung*) of the former fortifications land in the middle of the century.

The city gates remained a place of close public observation long after the gates were no longer closed at night. The *Leipziger Tageblatt* published daily "gate notices" (*Thorzettel*) listing the names of those admitted through the gates, along with their places of accommodation in the city. Because the notices were organized by gate and hour of arrival, reading them in the daily paper could give the impression of having "watched" at each of the gates. Leipzigers

1.3 This map of Leipzig in 1828 shows the promenades and gardens around the former fortifications land, including the English gardens and the Swan Pond to the east of the inner city, and the new oval-shaped Augustusplatz. *Plan von Leipzig, aufgenommen 1828*, hand-colored drawing, Stadtarchiv Leipzig, StadtAL RRA (F) Nr. 40.

used these gate notices to facilitate business with visitors, as can be seen in the newspaper's response to complaints that the failure always to include exact addresses made it difficult to find those who had arrived.[36] Although the place where visitors registered shifted from the inner to the outer gates in 1839, the practice of publishing this list of names in the daily paper continued otherwise unchanged until 1841. At this point, the list changed names from "gate notices" to "arriving strangers" (*einpassirte Fremde*) and ceased to be organized by gate of arrival, instead listing visitors entirely by place of accommodation.[37] Although the function of the list remained the same, these changes distanced the public listing of visitors' names from the physical geography of the city border. No longer did the daily notices evoke a landscape of fences and gates. The publication of daily "gate notices" demonstrates one of the ways in which the geography of sight on Leipzig's border served the conduct of business in the city, especially as related to the yearly markets. The promenades around the gates gained in visibility at market times, as visitors to the city were encouraged to "see the city" by walking around the promenades. The paper ceased to print a list of visitors altogether only after 1861.

Leipzig's defortification opened a transitional period when the primary goal of planning on the edge was to make the city appear open and inviting to wanted visitors but still closed to unwanted ones. The city's many borders were regulated by a complex geography of sight and sight lines: who would view the city? from what angles? who and what could be seen at the city edge? Like a good English garden, the open trade city was to have clear limits, but those borders were to be obscured from view as much as possible so as not to affront the eye with the dark and archaic signs of a closed space (see figure 1.3). Yet defortification, along with the laying out of public gardens and promenades on the former fortifications land, did form the basis for heightened expectations of openness, as demonstrated by the public demands in the following decades for freedom of movement in and out of the city. By the 1840s, even signs in the urban landscape that reminded viewers that the city ever had been closed in the past seemed to some residents an affront to the city's identity as an open, growing space of enlightened and cosmopolitan principles. Conflicts over the border of the city were played out not just over who would be let in or kept out but also over access to sight, visibility, and light on city edge spaces.

THE GEOGRAPHY OF SIGHT ON THE LEIPZIG RING

Designing the promenades as a communal space that knit together the city and its hinterland meant delineating (or obscuring) not only the borders of the city itself but also the borders between private and public space, between leisure time and work, between the city as a marketplace and as a civic community—and to do so through garden and landscape design. In this way aesthetic, political, and economic decisions were often inseparable. In debates over the proper shape to be taken by the promenades, Leipzig's city government and planners had to answer difficult questions about the meaning of natural resources and built landscapes for the city's social and economic order. Was sunlight more valuable for leisure or for labor? Did trees provide shade for strolling burghers? or did they steal necessary sunlight from productive craftsmen? Was the urban edge a space of public display and conspicuous consumption or one of poverty and charity? Whose walls and private gardens should be publicly viewable from the promenade? and whose were better off hidden from public view? The planned sight lines of the carefully conceived gardens along Leipzig's former fortifications framed private wealth as public good while obscuring from view the less desirable members of the urban community.

This translation of social and economic relations into aesthetic principles was not a straightforward process. Instead, it was contested from the late eighteenth century onward. Individual citizens deployed the language of urban improvement and progress to represent their private interests and to frame their relationship and contributions to the urban community. When the question was the hiding of undesirable housing and ramshackle gardens, the city council argued the necessity of giving the city a consistent and "orderly" border. When at issue instead was the revealing of private property for public view, the same walls and trees that seemed orderly in one context evoked the dark unfriendly gates of the past instead. Conflicts over nature in the promenades did not fit easily into an opposition of city versus nature. Instead, particular uses of nature were part of the urbanizing of space: garden planning and city planning inextricably connected on the urban horizon.

On the one hand, owners of grand villas adjacent to the promenades sought greater visibility for their houses from the gates and promenades, framing their private wealth as a public good. Shortly after the removal of the city wall began, the wealthy master carpenter Friedrich Christian Liederitz lamented that the large linden trees along the promenade obscured public view of the prettiest parts of his house. Gathering support from his neighbors, Liederitz's petitions to the city council are particularly revealing for the ways in which he combined questions of social status, public health, and aesthetics. He argued that the removal of trees along the busiest sections of the promenade would benefit public health by increasing the circulation of air in the city and hence protecting against fevers caused by cold and damp. In addition to improving the health of the city, he added, removing the trees would also improve the city's appearance and the delight of passing promenaders. Name checking some of the city's most magnificent villas, he made clear that this was also a question of social aspiration. He feared all the money he had spent in decorating would be wasted if his home remained hidden from view. Finally, in linking questions of urban improvement, aesthetics, and social status Liederitz made a historical argument, when he framed the opening up of sight lines at the urban edge as a continuation of the project begun in the earlier removal of the city wall. Would it not be inconsistent, he asked, to leave dark gloomy trees around the city after the council had done so much good by removing the gloomy city gates and walls that had been there before? While Liederitz wished his home to be publicly visible, petitions about sight lines in the promenades went the other way too. For example, in 1832, the residents of several dwellings near one of Leipzig's gates complained that they had been "deprived of their view into the promenades by the planting of many new trees" behind their houses, arguing that their ability to purchase these houses also gave them the right to this interesting view—presumably not just of gardens, but also of the promenaders themselves.[38]

At the same time, Leipzig's transformation of its fortified edge into a space of public display where the city and its proudest buildings and civic institutions were to be viewed—where wealthy burghers went to see and be seen—was repeatedly challenged by the presence along the wall of some of the city's less attractive and most impoverished housing as well. Of particular concern was the narrow space enclosed on one side by the high stone wall around the city and on the other by the lower, earthen wall that earlier had run along the city's moats. This was known as the *Zwinger*, or ward. It was a narrow strip of land that held a liminal place in Leipzig's urban geography, neither clearly inside nor definitively outside the inner city.[39] According to one city guide, the ward had earlier been "filled with miserable huts," but by 1792 these were replaced by little houses "rented out for a small fee to its own lower officials and other inhabitants," in some cases built directly against the inner city wall. By the Grimma Gate, a row of low buildings accommodated city soldiers. A nearby three-story structure housed students, in convenient proximity to the university. By Halle Gate the ward was inhabited by skilled workers and modest craftsmen, while to the south of the city by Peter's Gate the ward was populated by the city's very poorest residents.[40]

Hidden between two walls, this narrow strip of land had been dark, hidden, swampy, and out of the way, but the transformation of the fortifications land and the proximity of the new heavily trafficked promenades lent the space a new visibility. When several large trees were removed from along the avenue between Grimma Gate and St. George House in 1832, those taking a stroll along the promenade had a good view into the gardens and dwellings, creating, according to the city council, a "most unfavorable sight . . . aris[ing] in part from the most tasteless way in which the inhabitants of the ward have built and planted" the visible slope between the ward and the promenade.[41] The dispute hints at a deeper tension between Leipzig's bourgeois culture and that of the working-class inhabitants of the ward. The city council's communications on the subject bristle with disdain—repeatedly mentioning the visible laundry that caused "bitter remarks and mockery" among promenaders and reprimanding the ward residents for having no sense of "morals and order."[42]

Few descriptions exist of the more impoverished dwellings around Peter's Gate, but a lengthy article in the journal *Der Leuchtturm* (The lighthouse) by the prominent author and democrat Hermann Semmig in 1847 provides an account of the slums in the remaining depressions where the moats had once been and the difficulty they posed for the view of the respectable city presented along its edge. "Leipzig has long had its 'wild quarter' where filth, revolt against the social order, prostitution and so on flourish abundantly," wrote Semmig, but the neighborhood, popularly known as "Free Switzerland," had seemed to escape the attention of most Leipzigers, its neglect both result and cause of its characteristic topography. Semmig wrote: "When one goes from Peter's Gate along the promenade to the Augusteum, the view of the inner city below will be restricted by tall, attractive buildings and the imposing aspect of these will give the viewer a good impression of the grand city of Leipzig. But behind these high buildings is another Leipzig of dirty, of sick or even savage appearance."[43] Semmig directly links this "revolt against the social order" with political revolution, pointing out that the "Leipzig Massacre" of 1845 (when protests against the Saxon prince led by Robert Blum resulted in the death of eight Leipzigers) took place in the Roßplatz, directly adjacent. This area was among the very last sections of Leipzig's moats to be leveled: economic and social geography shaped physical topography.

Another conflict arose from the different use of space by different classes in the former fortifications land along the promenades, and this concerned not sight lines but sunlight. In 1832 inhabitants of the ward by Halle Gate asked that the large shade trees planted along the former city wall be removed because they deprived those living in the ward of the warmth and sunlight necessary to preserve their health and to engage in productive labor as carpenters, shoemakers, and craftsmen of other sorts.[44] The dappled light that filtered down through the leaves was distracting to their labor. The comfort of Sunday strollers was at odds with the virtuous industry of those who had to work both to support their families and to pay their taxes. The city council rejected their request, reminding them how much they had already benefited from the removal of the city wall and reasserting the importance of protecting promenaders from oppressive summer heat.

What did the trees in the promenade signify? Depending on the viewer, the trees that lined Leipzig's promenades were either a reminder of the dark walls they had replaced or a symbol of Leipzig's open and prosperous future, a reminder of how far the city had come in the last decades. They were, of course, both. Although the "civil conversion" of the city border gave expression to Leipzig's *bürgerlich* identity, a closer examination of the history of the promenades reveals the degree to which this identity had to be actively defended—sometimes tree by tree. The view of the city in the panoramic stroll emerged out of conflicts over sight, light, and the use of public space on the urban edge.

LEVELING THE GROUND AND REMOVING REMINDERS OF THE PAST

Entering the second half of the nineteenth century, Leipzig became one of Germany's fastest growing cities. It counted only 35,230 inhabitants in 1811, but its population would pass 100,000 in the year of German unification in 1871. Leipzig had long been a commercial center with special market privileges, but from the 1830s onward industry fueled the growth of its suburbs. In 1833 Friedrich List inspired Leipzig's businessmen to petition the Saxon government for the opening of a new railroad, and in 1837 the first railway journey on one of Germany's first railway lines was completed between Leipzig and Dresden—with a train suitably dubbed "Faust"—opening a period of rapid growth and construction.[45]

This changed how Leipzigers thought about both the promenades and Leipzig's other physical and administrative borders. Originally, Leipzig's promenades signified openness and modernity. As part of the "civil conversion process" of the city's fortifications, the promenades encapsulated Leipzig's identity as an open trade city. But over time, the meaning of Leipzig's promenades shifted within the changing landscape of the city. As other markers of the urban edge fell away, the promenades became more clearly a marker of the city's past. Increasingly, observers commented that the promenades recalled the enclosure of the old fortifications, and the limitations on growth and movement they had once imposed—instead of evoking the process of their removal and the subsequent opening of the city. For some, this was a positive association, indicative of continuity in Leipzig's urban identity. For others, the promenades themselves came to be seen as an arbitrary restriction placed on the city's economic and physical growth. These conflicts heightened as the density of settlement in the inner city and traffic through and across the promenades increased. An editorial in the *Leipziger Tageblatt und Anzeiger* (The Leipzig daily news and advertiser) published in 1844 called for the removal of trees in the promenade, because the plantings' original purpose had been to obscure the visual affront of ugly and crumbling city walls. Now that the wall had been removed and the old city edge was lined by attractive houses, the trees no longer served any purpose. The park, the anonymous editorialist wrote, was "colliding with the city" (*in Collision mit der Stadt*).[46] Where did trees belong? Clearly not amid the traffic and bustle of an urban space.

Complaints that the promenades were out of keeping with the character and needs of a busy commercial city like Leipzig appealed to arguments for a kind of inner colonization of the city itself, in which municipal parks were seen as unexploited space within the city's borders. A particularly forceful and detailed petition for "an expansion of the *inner* city" came in 1841 from businessman Daniel Sellier, who was also a founding member of the Leipzig-Dresden Railway Company. Sellier drew up a detailed plan to replace a large stretch of the treasured green space with new streets and market spaces. His proposal received a chilly reception from Leipzig's building director, Albert Geutenbrück, who defended the promenades by appealing to their historical origin in the old wall circuit. He wrote that Sellier's proposal "would only destroy what was built up over centuries."[47] Both Sellier and Geutenbrück depicted the promenades as a symbolic preservation of the old form of the city, repicturing the trees, gardens, and paths in the historical context of the old fortifications. One thought such remnants of an outgrown city identity had no place in the urban landscape; the other celebrated them as part of the city's heritage.

These conversations were made all the more complex because spaces of leisure and commerce often overlapped at the urban edge. Market booths lined the promenades during market times, and private gardens also provided overflow space for weekly and yearly markets. In the middle of the century, Lehmann's Garden (formerly Kleinbosische Garten) outside the Little Barfüß Gate acquired the nickname "Leipziger San Francisco" for the jumble of provisional market booths that would be set up there during market times.[48] Hence, planning the Leipzig Ring and leveling the old moats and bastions had economic as well as social and aesthetic implications. The land to be leveled was not empty but, rather, filled with market booths and workshops. Concerns about what would happen to these market spaces played a central role in the lively public debates of the 1850s over the filling of the remaining moats around Peter's Gate. In an opinion published in the *Leipziger Tageblatt*, for example, one advocate of preserving the moat as it was, wondered: "if the moats are brought to ground level, / where will the rope makers go?" A second asked "where will the [market] *booths* go?" speculating that only when all the small booths had disappeared from the landscape entirely would the project of "leveling" (*planiren*) be complete. A similar set of concerns attended discussions in the 1850s over the market booths in Augustusplatz and whether or not to level the Roßplatz, both important open spaces along the promenade in the former fortifications land.[49]

What these discussions show is that conflicts over the "leveling" of the city by the forces of capital, which Rudy Koshar has documented in the beginnings of the urban preservation movement in the 1890s, actually originated much earlier in discussions of plans for former fortifications land in the first half of the nineteenth century.[50] Since former moats and the spaces along city walls had long been used by craftsmen for labor and by the proprietors of small market booths for which there was no space in the main marketplaces, leveling the city quite literally removed the spaces for small craftsmen and businesses and raised concerns that in the modern city there would be no room for smaller concerns to compete in a rationalized, leveled, marketplace. The relationship between insiders and outsiders in the city was complicated by Leipzig's significance as a market city, as a university city, and as a center of the book trade. Even more so than most other cities, modern Leipzig was a city defined by its openness—to goods, to people, and to ideas. This meant that managing Leipzig's borders always raised the question as to what an open horizon should look like. How should the edge of the growing city look? Should it even be visibly marked? In the middle decades of the nineteenth century, attention began to turn from the promenades and former fortifications land to the space of the suburbs beyond.

VISIONS OF GROWTH IN THE SUBURBS

An 1850 depiction of Leipzig helps to visualize a dramatic change in the city's identity as a space of growth when contrasted with earlier bird's-eye views of the city as a static, walled space (see figure 1.4). The dominant visual elements of this image are roads and spaciousness. The city fades into countryside, and the viewer is invited to imagine the city's future development. This reads as a picture not just of a city as it is in the present but also of the movement of the city as it expands. Yet even in this era of expansion and growth, the city's boundaries in their appearance and function remained a source of anxiety. In addition to ongoing conflicts over the appearance and use of the promenades and former fortifications land along the Leipzig Ring, the appearance of Leipzig's suburbs and outer boundaries became another source of conflict in the middle decades of the century. Conversations about the suburban border reiterated many of the themes of conflicts over the city's inner boundaries. Leipzigers often thought of the outer border of the city in relation to the in-

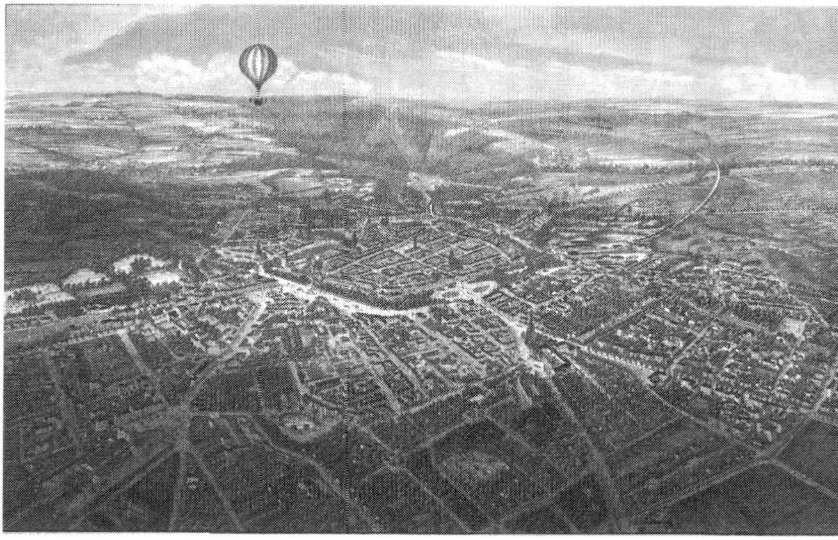

1.4 A bird's-eye view of Leipzig looking north toward the Rosenthal Forest invites the viewer to imagine the city's future growth. Here there is no clear boundary between city and nature, center and periphery. Adolph Eltzner, *Leipzig*, 1850, engraving, 43x28cm, Sächsische Landesbibliothek— Staats- und Universitätsbibliothek Dresden/Deutsche Fotothek. http://deutschefotothek.de/documents/obj/70300824/df_dk_0002833/.

ner walls and vanished fortifications, even though the outer border was very different in both function and appearance. The wall of the suburb was never available as a visual boundary to the city, in the form of a panorama or the circuit of an afternoon stroll. Instead, it was lost in the distant horizon of a bird's-eye view, overcome by anticipated growth.

One necessary step in this shifting understanding of the edge of the city was the elimination of the remaining administrative distinctions between the city within the ring and the outer suburbs on its other side. The inner-city wall had remained a kind of border after 1824 in the availability of certain urban services, especially nighttime street lighting. The first steps were made to extend street lighting to the suburbs in 1830, but progress was slow and uneven.[51] In petitioning for urban services like street lighting, suburban residents referred to the recent removal of the city wall as justification for their claim: if this physical marker of the old urban edge had been removed, why should that border continue to function in other ways? An 1835 petition signed by over two hundred suburban residents justified their claim to street lighting, both appealing to the need for a united, integrated city and referring to one of the last and most visible pieces of the city wall to fall:

> Taken away are those old monuments to earlier, sadder times, and with the Grimma Tower the last aspect of those times falls into complete oblivion. Leipzig resembles a garden: each part is in happy accord with the other in an orderly interrelation and it displays such an inner consolidation, such a tight connection between city and suburb, that one can rightfully say, "There is only one Leipzig." But, unfortunately, this is only true during the day, and in no way can one call this out in the evening or at night. Then, upon leaving the nicely lit city and the just as brightly lit promenades and entering the streets of the suburbs one finds complete darkness.[52]

The residents' description of the city as a garden linked their appeal to the promenades that surrounded the inner city and that had become Leipzig's proudest asset. The authors went on to point out that lighting the roads leading to Leipzig, an important trade city, would create benefits beyond those enjoyed by local house owners, but they also noted that, since they paid the same taxes and suffered the same burdens as other burghers, they ought to enjoy the same benefits. Similar petitions came from Leipzig's other suburban neighborhoods. The residents of the area outside of Peter's Gate had petitioned for the provision of street lanterns in 1829. Inhabitants of the St. Johannis suburb submitted another petition in 1832, complaining that the darkness in the suburbs made it difficult for midwives to find their way at night.[53] In response to the 1835 petition, a member of the city council issued a detailed report examining the historical relationship between Leipzig and its suburbs and debated whether a full equalization (*Gleichstellung*) of the two was called for. The city council concluded that the only reason for any difference between inner city and suburb lay in the old fortifications, and hence the distinction no longer held. In 1839 Leipzig's suburbs were fully equalized with the city, with a unified tax code and access to urban services such as street lighting.[54]

As a result of this change, the suburban borders gained in significance. The outer gates had already been the location where excise taxes were exacted, and they had in 1830 become an important point of defense against urban unrest. The outer border, with its twelve gated points of entry, was now also the edge of the city itself. Unlike the promenades or erstwhile fortifications, there was nothing grand about this outer border. Until 1840 it was not even consistently marked by a wall. Most of the gates consisted of iron grill, flanked by stone or wooden pillars (see figure 1.5). After the incorporation of the suburbs, a modest fence was built along the municipal boundary. The tight control of movement in and out of the city was, however, a matter of

1.5 This drawing of one of the outer gates of Leipzig was provided to the city council by Carl Lampe in 1836 in support of his petition to widen the street. Stadtarchiv Leipzig, StadtALTit. XXIVc Nr. 61, f. 57.

much concern to the city council in order to prevent tax fraud. In the 1840s the Dresden Gate was moved further out, to the municipal border, since it was now more important that the point of control be on the edge.[55]

This construction of a new city border was out of tune with the spirit of the times. The new suburban fence resulted in the so-called boarding up (*Verplankung*) debate over the cost of clearly marking the urban edge in a decade of trains and revolutions. In May 1843 the *Sächsische Vaterlands-Blätter* (Saxon Fatherland's paper), a liberal journal under the direction of Robert Friese (an associate of Robert Blum's), published an indignant editorial complaining about Leipzig's new border fence: "The world is not boarded up [*mit Bretern vernagelt*], but Leipzig is to be. There may well be people whose insight is boarded up, but there are also many who would rather not allow either their insight or their outlook be boarded shut, and among those are especially those wise inhabitants of the outer border of this admirable and loyal city."[56] In this editorial, the owners of property along the city's border are depicted as valiant pioneers. Those who lived and owned property along the new outer wall had apparently complained about the inconvenience to city authorities.[57] Highlighting the hubris of any inevitably futile attempt to enclose the growing city, the author predicted that in response to their complaints "this new Great Wall [*chinesische Mauer*] [would] soon disappear again." His outsized analogy links the struggle of a handful of Leipzigers against the unpleasant encroachment of a fence's shadow into their private gardens with a global struggle against irrational borders. The author expresses particular indignation that at a time when Germans were engaged in the business of tearing down walls, the city of Leipzig was engaged in the untimely business of erecting new ones.[58]

Both before and after 1839, the outer border and the control at its gates were cause for resentment among those suburban residents who were most affected. Only a decade after the gate closure at the inner gates had been lifted, they reprised many of the same complaints about freedom of movement and unequal taxing. There were petitions throughout the 1830s for easier passage through the outer gates, which were closed at 10 or 10:30 at night. Residents complained of inconvenience and of the rudeness and drunkenness of the gatekeepers, a grievance that highlighted the sense that such restriction of movement was an arbitrary and irrational exercise of power.[59] In 1835, the city council received a petition from one of its own on behalf of those who lived outside of Windmühlen Gate, pointing to the recent rapid development of the area around the gate as justification for the extension of its hours. This was followed by petition in 1838 for the rights of those living outside of Windmühlen Gate to be able to pass through at any time of day or night without paying a fee.[60] Especially in the 1850s, increased traffic led to more accidents in the narrowed passage through the gates. There is some evidence of passive resistance to the gate check. In 1859 the gatekeeper at Windmühlen Gate complained that those riding inside carriages entering the city at night sometimes refused to give their names.[61]

Saxony's entry into the Zollverein (Customs Union) in 1834 had mixed consequences for Leipzig's importance as a commercial center. Although the city's trade fairs continued to grow in the decades following, they did not keep pace with leading industrial sectors, as industrial products not easily sold at traditional trade fairs gained in significance. The Leipzig fairs maintained their significance after 1860 by changing from traditional "goods fairs" (*Warenmessen*) to modern "sample fairs" (*Mustermessen*), in which models were displayed for later order and delivery instead of purchase, a shift that also diminished the importance of controlling the flow of goods in and out of Leipzig's urban space. In 1861, the Leipzig city council let the market privilege (*Marktrecht*) that necessitated the collection of excise taxes at the city border expire with the year. Trade guilds were abolished in the same year. The outer wall now lost its final significance, and the gates fell into disuse.[62]

If the planks of the suburban fence that marked the old edge of Leipzig's market privilege was so out of keeping with the character of the city and the appropriate appearance of the urban edge, what should that periphery look like? From the 1830s on, a series of new projects for the intersection of nature, industry, and recreation emerged in Leipzig and elsewhere to give new form to the idea of the urban edge as a space of growth, in the suburbs and beyond. The boom years of growth in Leipzig following the middle of the century were characterized by ambitious plans for industrial suburbs and "frontier towns" conceived by a new kind of urban developer. One of the most influential developers in Leipzig was Carl Heine (1819–1888), who had his career roots in urban water control and planning but who later worked on large-scale transformations of the Leipzig hinterland for both industry and residence, especially in the industrial boomtown of Plagwitz. The national and local press described his developments in comparison to the rapid settlement of the American West.[63]

As the city edge became defined as a space of growth and potential expansion, Germans considered access to nature from the city in new ways. In the 1850s and 1860s, two Leipzig professors conceived contrasting models for the individual's relationship to nature that would have lasting significance for the landscape of the German city and its hinterland. Their projects involved, respectively, two of the most characteristically German of peripheral urban green spaces in the industrial age: the Kleingarten (or small garden allotment) and the city forest.

The first was championed by Daniel Gottlob Moritz Schreber (1808–1861), a medical doctor born and trained in Leipzig. Early active in the Turner movement as well as in city politics, Schreber took up public health education in the final years of his life. He worked along with the director of Leipzig's Bürgerschule, Ernst Innozenz Hauschild (1808–1866), on the access of city children to green space as a way of righting a host of health problems presumed to be a consequence of the cramped and impoverished conditions of modern urban life. Although he did not himself live to witness the result of these efforts, he lent his name posthumously to the Schrebergarten movement through Hauschild's founding of the Schreber-Verein in 1864. In his medical guide to indoor gymnastics, Schreber presented gardening as the perfect physical activity for maintaining health and balance in one's mental and physical beings. The artificial movements of medical gymnastics were only a compensation, or medical remedy for damage already done, not a comprehensive solution to the sedentary life of the average city dweller. The provision of garden space, then, was by contrast a preventative measure, making space for natural, whole-body activity with easy access to the city, with its focus on young children, their schools, and their families.[64] Under Hauschild's direction, and following Schreber's principles, the Schreber-Verein set up a playground for children surrounded by allotment gardens, first for flowers and later edible plants, called "family gardens," which were opened not just to children and their parents but to other residents of the city as well. Leipzig's garden settlements were among the earliest in Germany.[65]

As Schreber and Hauschild worked on their family gardens, Adolf Emil Roßmäßler (1806–1867), another Leipzig academic, worked on his own plan to reform access to nature on the city edge. Likewise concerned with the health of growing urban populations in the working and middle classes, Roßmäßler envisioned not a garden on the urban edge but a forest. To develop individualism and connection to nature, his ideal exercise was not planting vegetables and pruning fruit trees but a hike through the forest. Roßmäßler was an early advocate of forests as socially equalizing spaces. He was also politically active, and one of the cofounders of Leipzig's first workers' education societies in 1860. Combining his interest in nature with engagement in the city's politics, he argued that the working classes should have access to nature. In Leipzig, access to the forest meant the Rosenthal Forest, a royal hunting preserve that turned into an urban retreat and bourgeois pleasure ground in the middle years of the nineteenth century. One of the distinctive characteristics of the German forest seems to be its accessibility from the city. As Jeffrey K. Wilson explores in his study of the German forest, botanists and forest enthusiasts imagined a German national landscape connected by accessible forests, in which city dwellers across the country would have access to weekend hikes in the woods.[66] In a way, Roßmäßler and others imagined the forest as a common horizon for the edge of the German city.

Schreber and Roßmäßler, influenced by the politics of the 1840s, shared a common concern with working- and middle-class urban bodies and how those bodies were shaped by their relationships to the natural world. At the same time that advocates of urban business and growth argued that trees and promenades did not belong in the city itself, Schreber and Roßmäßler offered new models for green space on the urban periphery. They provided a vision for the urban edge that allowed a reconciliation of growth and access to nature, envisioning the urban horizon as the interface between the city and nature in ways that also proved influential for the rest of Germany. Later in the century, Saxony became one of the centers of the *Heimatschutz* (homeland preservation) movement. Roßmäßler's ideas about the German forest would later be picked up by the Heimat movement in Saxony, pushing for diverse and interesting forested landscapes against a feared transition to timbered monocultures.[67]

As a way of viewing the city, one effect of the street-line panoramas was to free the city at its edge from any relationship to its immediate hinterland. In these promenade and ring-road panoramas, the viewer's back is always turned to whatever lies beyond the promenades. Placed in relationship to its own history but not its geographic location, the city effectively floats in space, available for viewership and consumption by visitors from abroad but not connected to the land around it. In the context of this floating space, the presence of nature became problematic and conflicted—trees at odds with images of growth, garden spaces subject to visions of "inner urbanization," and so on. The projects of Roßmäßler and Schreber addressed a renewed need to find an accessible space outside the city and to define a common horizon of urban space not defined by a wall or border.

As their city's borders changed, Leipzigers learned to view their city in relation to its past. City planning often functioned as a set of strategies for manipulating sight lines along with traffic flow. This manipulation of sight could be both hyperlocal, as in what views do trees conveniently hide? who and what most deserve access to sunlight? and then also global, as in how does the appearance of the city reveal its place in the world? what distant places do local landscapes evoke? what global processes are revealed in urban spaces? when does a simple garden fence conjure the Great Wall of China? Constructing the Leipzig Ring as a representative urban space in a panoramic journey around the city's edge was a process that took decades of planning and negotiation. That process involved hiding some spaces and closing them off as well as opening up other spaces. In the period after the city wall lost its function as a military and policing border for the city the idea of the open border was manipulated both to reinforce and hide social and economic boundaries and also to decide what "views" were to be assembled into the patchwork image of the urban panorama.

Over the course of the nineteenth century, urban growth itself became something to be watched, visualized, and presented to the urban public.[68] By the middle of the century Leipzigers began considering the edge of the city in a new way, constructing urban horizons that made visible not just the bounds and definition of urban space and community but also the ongoing growth of the city in frontier settlements and cultivated green spaces that addressed the social needs of new and growing urban populations. Yet, even as those new horizons of growth opened up, the old borders of the city remained important touchstones for Leipzig's urban identity, linking the present city both to its own past selves and to the process of growth and development. As the city edge changed, Leipzig also retained its identity as a historically rooted community—in part through the ways in which Leipzigers continued to repicture the changing urban landscape in the context of its historical boundaries.

2 Conquering the Wasteland

Oldenburg's Urban Empire in the Northwestern Moors

In 1924 a representative in Oldenburg's legislative assembly accused the mayor of the small German state's eponymous capital of megalomaniacal plans for urban expansion. "Macedonia is too small for you," he charged, "seek out another Kingdom." The action that provoked this bitter allusion to Alexander the Great was the city's most recent attempt to annex a large neighboring district. Oldenburg's enterprising mayor, Theodor Goerlitz, envisioned a great economic future for the city as a center of industry and commerce, for which spatial expansion was, he thought, a necessary prerequisite. Where residents of Oldenburg's hinterland saw outsized imperial ambitions, Mayor Goerlitz saw the righting of a historical shortsightedness that had long constrained Oldenburg's healthy and natural growth within artificially narrow borders.[1]

While the conditions of the Weimar era city were new, the debate over the proper limits of city expansion in the 1920s recalled a long history of controversy over the nature of Oldenburg's frontiers. City administrators had been eyeing much of the land incorporated in the 1920s since the middle of the previous century. Ever since the city edge was no longer visibly inscribed in the landscape by the stone, earth, and water of its premodern walls, ramparts, and moats, Oldenburgers had struggled to find a new basis for defining the appropriate limits of the city and so to demarcate the shifting economic, environmental, and political relationships among city, state, and hinterland.

Oldenburg might seem an unlikely site for accusations of urban imperialism, the ambition of growing cities to conquer ever widening hinterlands through a combination of settlement, environmental transformation, urban infrastructure, cultural and economic assimilation, and legal annexation. The sleepy ducal seat has never been among Germany's greatest cities. Unlike Leipzig, the city until relatively late remained sheltered from the storms of the industrial revolution. The revolutions of 1848 were here a distant echo. The spreading network of railways only caught the city in its web in 1867, by which time the city had reached a population of about fourteen thousand. It would not attain the status of a "big city" (*Großstadt*) until after the Second World War, when it received a flood of refugees and became, as one author called it, a "big city against its will."[2]

The very features that made Oldenburg seem an unlikely place for grand metropolitan ambitions—its late growth and secluded location—clarified the city's imperial relationship to its hinterland. Key to nineteenth-century urbanism was a claim to universal applicability as a way of organizing both space and people, divorced from the particularities of local environments and populations, but those urban universalist claims seemed out of place in such a provincial capital as Oldenburg.[3] When Oldenburg's administrators and boosters planned for the city's growth, they combined appeals to its distinctive history as an important regional capital on the one hand and an expectation of universal urban growth on the other, even though these appeals were often in tension with each other.

Nineteenth-century assertions of urban universalism combined a cultural claim—that cities shared common ways of socializing, working, and playing across regions—with a technocratic claim resting on confidence in the ability of urban infrastructural technologies to urbanize peoples and natures of different kinds. Urban expansion in Oldenburg was also what David Blackbourn calls a "conquest of nature" and, particularly, a conquest of the landscape of water.[4] The regulation of borders was invested with special significance in the German Northwest, where the boundaries between water and dry land were both constantly changing and often a matter of life and death. In the moor, marsh, and geest of the Northwest, political and economic might translated into the power to control and transform the landscape, to hold back the flood, to canalize the moors, and to make land where once there was amphibious marsh. Power struggles between the state of Oldenburg and its neighbors played out through competing projects to improve the wastelands of northwestern Germany. Similarly, power struggles over the expansion of the city of Oldenburg engaged questions of landscape, water control, and improvement.

In Leipzig the dynamic of urban expansion was governed by ideas of light, sight, and the marketplace, but in Oldenburg urban expansion became primarily a question of power and land. Oldenburg's premodern urban borders were relics of both water control and political power. The shifting boundaries between sand, mud, and floodwater provided powerful metaphors for urban transformation, grounding social and political change in natural landscapes and environmental discourses. While a number of North American urban historians have highlighted the ways in which urban expansion intersected with environmental conquest and transformation, these intersections have been less well developed in studies of German urban landscapes.

Environmental historians such as Thomas Lekan and William Rollins have identified the German relationship to nature as providing a possible middle way between unconstrained industrial intervention in the environment and the sanctifying of nature as an inviolable wilderness untouched by humans.[5] German Heimat instead embeds human communities within natural environments, with cultural and nature conservancy being two sides of one coin. Likewise detailing the connections between culture and nature, Blackbourn's work on waterscapes recounts the longue durée of German history through Germans' successive transformations of the land. The inscription of national community in nature provided an illusion of environmental stability as the transformations wrought by one generation in the name of progress became the unspoiled natural landscapes of the next. Cities underwent the same process of transformation described by Blackbourn, as urban infrastructure itself became subsumed over time into the natural landscape. Urban border systems became obsolete forms of urban infrastructure over the long nineteenth century, and contemporaries went from seeing urban walls, moats, and borders as pieces of infrastructural technology controlling and changing the landscape to instead subsuming historical urban borders into their understanding of the landscape itself.[6]

THE GEOGRAPHY OF WATER AND DEFENSE

The city of Oldenburg sits south of the Jade Bay at the junction of the Haaren and the Hunte Rivers. The Hunte is a tributary of the Weser, connecting Oldenburg to its powerful Hanseatic neighbor Bremen about fifty kilometers to the east. A tongue of rich green marshland stretches out along the Weser River north to the Jade Bay. To the south and southwest spreads the fenland of the Hunte-Leda glacial valley. The city sits on a southern-reaching spur of the Oldenburg/East Frisian Geest—high, sandy, barren terrain left behind by retreating glaciers.[7] The region's landscape is shaped by shifting borders between dry infertile geest, desolate boggy moors saturated by groundwater, and fertile coastal marshes—islands of productive land in a largely bleak environment.

Oldenburg benefited early from a favored position with regional transportation on both land and water. Pinched in by unproductive high moors to the west and moor and marsh to the east, the geest on which it sits provided solid ground for road construction. The busy high road linking Frisia with Bremen channeled traffic through the city, while the Hunte River provided a navigable waterway to the Weser here where the Haaren joins its course and widens its bed. A regional transportation hub on the border of different producing regions, Oldenburg has served throughout its history as a "bridge city."[8]

If roads and waterways invited people in, Oldenburg's location also provided the second important ingredient for a premodern city—ideal conditions for shutting unwanted visitors out. The old city core established atop the natural elevation created by deposits from the Haaren and the Hunte enjoyed conditions well suited to the construction of a fortress. The city gained natural defenses from the wide floodplain of the Haaren River—called the Dobben—to the west, which flooded each fall and spring, and the Hunte's floodplain to the east. Controlling the rivers meant that much of the area around the city could be flooded as part of its defense. Only in the dry open geest to the north did the city lack natural protection or possess easy ground for later expansions.[9] Although it had already been a site of settlement for centuries, the story of Oldenburg's organized development began in the twelfth century when Count Egilmar II took advantage of quickening trade and the area's naturally sheltered position to construct a *Burg* in the bend of the Hunte, giving the site its first fortifications. The New City north of the castle prospered, and the original walls were extended in the early fourteenth century around the time when the city received a constitution on the model of Bremen's.[10]

From the early sixteenth century, when it gained status as a fortress city, through the middle of the eighteenth century, the city of Oldenburg developed an increasingly elaborate system of military fortifications that both separated the city from its hinterland and represented the first major expansion of the city.[11] As the city's needs for defense, space, and navigable waterways increased, the settlement became a decisive factor in shaping the surrounding land. Oldenburg's growth occurred only as a result of a radical transformation of its hinterland. This was true not only of the land reclamation projects that accompanied the city's nineteenth-century expansions but also of an earlier era characterized by the urban geography of defense. In 1667 Oldenburg lost its independence and became a province of the Danish Kingdom.[12] Oldenburg suffered economically and physically as a result, but the military importance of its capital city increased as one of Denmark's principal southern fortifications. Two expansions of the city's defenses were carried out under Danish rule, first in the 1670s and then between 1732 and 1746. Over the course of these changes, the Hunte was deepened, extensive excavation was required as part of the expanded fortification system, and the flow of the Haaren was redirected and its marshy stretches deepened to provide a steadier water supply for the city's moats.[13] These changes were controversial. Residents resented the expansions as an economic burden and a disruption of the landscape. They feared the loss of beloved "gardens and fields" around the city, and with them the city's unique character and appearance. In 1672 the city refused to supply labor for the construction of four new bastions, arguing that Oldenburg had been too weakened by poverty and famine.[14]

These expansive fortifications were the city's first extension of the built landscape into the marshy ground of its immediate hinterland. The wide band of defensive structures and the elaborate tangle of waterways at the urban edge exaggerated the break between the city and its surroundings (see figure 2.1). Furthermore, their design occasioned early reflections on the relationship between the shape of Oldenburg's settlement and that of the ideal-typical city.[15] Attempts to adapt the geometrical ideals of fortifications science and aesthetics to Oldenburg's topographical peculiarities prefigured later nineteenth-century concerns about the asymmetrical growth of the city occasioned by its terrain.[16]

The Danish king began to consider the removal of Oldenburg's fortifications less than twenty years after their expansion had been completed. Maintaining the elaborate structures was expensive and doubts had arisen as to the need for "defensible space" in the region.[17] Oldenburg's city government had no objections to the removal of the outer fortifications. Cumbersome and built for the purpose of defending the Danish Kingdom, they had little significance for Oldenburg's civic identity. In removing these, the city's main concern was the preservation of the water control systems that kept city and hinterland delicately balanced between flood and drought. Oldenburgers feared that the removal might result in the blockage and flooding of water channels around the city. The older inner fortifications were another matter. These

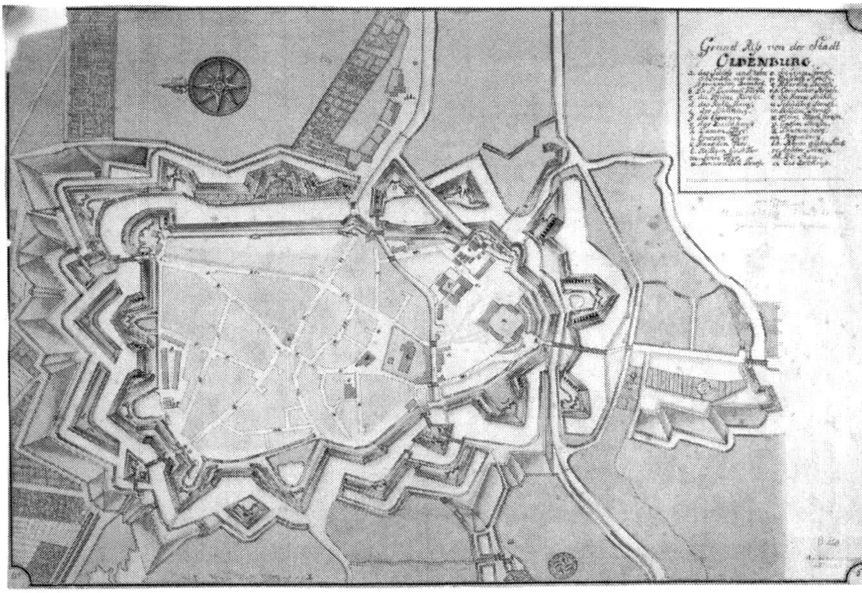

2.1 Oldenburg's elaborate fortifications, depicted here at their greatest extent, formed a dramatic boundary between the city and its hinterland. *Grundriss von der Stadt Oldenburg*, 1774, Niedersächsisches Landesarchiv Oldenburg, NLA OL Best. 298 OL B Nr. 120.

Oldenburg's mayor deemed "completely indispensable" to the city's identity, signifying Oldenburg's status, historical importance, and role as an administrative center. With its grand history, treasury, and archives, he wrote, "this is a place that one would certainly not like to make look like a village." In the end, the city got its way. The outer fortifications were auctioned off, and the inner wall and moats were sold to the city, which assumed financial responsibility for their upkeep.[18]

The expansion and subsequent removal of Oldenburg's outer fortifications changed the contact zone between the city and its immediate hinterland but did little to reshape the city itself. By the latter half of the eighteenth century, however, the city's inner ramparts had come to seem both costly and obsolete. By the time the removal of Oldenburg's inner fortifications began in 1789, the city's status had changed. Danish rule came to an end in 1773, and the County of Oldenburg and Delmenhorst was elevated to the status of duchy in 1774. The city of Oldenburg regained its status as a *Residenzstadt* when Duke Peter Friedrich Ludwig made it his capital in 1785. This marked the end of what had been a difficult period in Oldenburg's history. Duke Ludwig would oversee the city's economic and cultural revival, a rebirth many saw reflected in the remaking of its built environment. The removal of the old inner fortifications between 1789 and 1840 became one of the largest and most visible public works projects associated with the city's renaissance under the enlightened rule of Peter Friedrich and his successor the Grand Duke Paul Friedrich August.[19] The loss of the city's inner fortifications meant something entirely different now that the city was once again the capital of an independent Oldenburg.

Replacing the inner fortifications with promenades served several purposes simultaneously for both the city and Oldenburg's duke. When Duke Peter ordered the first section of the fortifications removed in the 1790s, he hoped to secure the city from future military occupation by making it a less desirable stronghold. He was also reacting to local unrest and economic hardship, defusing local tensions by demonstrating his largesse to the city. In the project's second stage, removing the ramparts to the north of the city in 1800, Peter expressed hope that the labor of demolition would bring bread to the bare shelves of the city's poor, likening it to a "second harvest" in the cold winter months.[20] Hence, the ducal "gift" of defortification to the city served the dual purposes of warding off danger from both without and within. Another important motivation for the city's defortification and the construction of urban promenades was, as Ewald Gäßler has argued, "to make as good and modern an impression as possible on those approaching the city from the outside." Peter undertook the paving of the roads leading out of the city and connecting it with its hinterland.[21] Open squares were built outside several of the gates, and the stone gates were removed and replaced with less imposing entrances. Stretches of the city moats were filled with earth from the dismantled ramparts and a path for public use was laid alongside adjacent private gardens. The promenades' circuit around the city became the pride of the city and the most visible mark of Peter's enlightened reign. Oldenburgers saw the linden-lined promenades and gardens along the city edge as compensation for the city's poor surrounding environment: here nature was controlled and improved, meant for enjoyment, in an area where the landscape was as yet little valued for its beauty.[22]

In the nineteenth century, Oldenburg's geography of defense increasingly gave way to plans for growth. When the city had been a fortress, Oldenburgers altered their city's hinterland to exaggerate its defensive advantages: channels were deepened and rivers redirected to exaggerate the trace of the moats. Later, these channels were diminished when the entering water was blocked, reducing the city's visible boundaries. This process continued for decades alongside the transformation of Oldenburg's hinterland with the regulation of new canals and the straightening and redirecting of both the Haaren and the Hunte Rivers. But even as the city expanded significantly beyond the old ramparts and moats, the former fortifications land remained central to Oldenburgers' perception of their capital, its status as a city, and the relationship between city and hinterland.

The city's history as a fortified place left a legacy of "military urbanism" that was evident in later planners' aesthetic preferences for regular, geometric forms.[23] Long after the ramparts were leveled, the fortifications land remained a reference point for the city's identity.[24] Into the twentieth century, the city's walls and fortifications played a key role in narratives of urban development, alternately representing the city's isolation and stunted growth and embodying the dense connections between the city and its surroundings. Even in pragmatic decisions of urban planning and infrastructure, the historical narrative continued to play a role in justifying decisions made by city builders into the twentieth century.[25] Early historians of Oldenburg, as of other cities, have often reproduced an idealized and simplified picture of the moment at which the city tore down its walls and fortifications, "burst[ing] its constraining chains and achiev[ing] space, air and light." In Paris, there was revolution, and in Oldenburg the city walls came down.[26] But this image of the city's expansion does not take into account how urban defenses, built up over centuries, had been integrally embedded in the environmental, as well as social and economic, relations between Oldenburg and its hinterland.

OPENING THE GATES TO GROWTH

As city administrators came to see Oldenburg not just as a ducal residence but also as a modern growing city with national and global economic connections, they applied to their plans an understanding of urban geography and identity that was divorced from the regional landscape. An important turning point in Oldenburg's development was the decision to open the city's gates in 1845, which set the stage for a conflict in the following decade between the mostly rural hinterland and the growing city. In supporting their desire to eliminate the nighttime gate tax, the city magistracy grounded their argument to the public in a rejection of the city's past as an urban fortress: the gate closure was rooted in the geography of defense, no longer suitable for an open and modern city.

Both before and after the removal of Oldenburg's outer fortifications, the city could be entered only by one of its five city gates. The Haaren and Heiligengeist Gates to the north, the Stau and Eversten Gates to the east and west, and the Damm Gate to the south. The points of contact between city and hinterland remained very stable during the course of defortification. Until the 1850s, few new gates or streets broke through the walls to accommodate increased traffic. As in many German cities in this period, the gates were closed at dusk, after which one could pass through only by paying the gatekeeper a fee, which was doubled after eleven or twelve at night.[27]

The gate taxes were a source of profit but, as in Leipzig, the closed city continued to serve a social purpose as well as an economic one. Controlling the points of entry to the city allowed control of impoverished travelers and beggars, whom gatekeepers were told not to let back in. (This was not very effective since such travelers frequently managed to pass for local farmers simply by leaving their luggage behind in a nearby inn.) The city edge had long been identified with poverty. One of the eastern bastions was known as the "wretched shacks" (*Elenden Buden*) and the so-called Gast Rampart was named for a "guest house," which housed the sick and the impoverished.[28]

For many Oldenburgers the rhythms of daily life required frequent passage in and out of the city. Many city residents maintained farms or gardens beyond the gates, especially to the north and northeast. In 1814 several city aldermen petitioned the Grand Duke to keep the city gates open later in the evening on behalf of city dwellers who required evening access to their gardens outside of the city. In the 1820s there were complaints about the gates' late morning opening times on behalf of the day laborers who lived outside the city and traveled through the gate on the Damm every morning for work (see figure 2.2). With the promenades and other new developments along the former ramparts, traffic after dark increased to beer gardens and other "places of amusement" outside the gates. The nighttime closure of Oldenburg's city gates gained importance with settlement spreading beyond the moats. Although small numbers of houses had grouped along the roads outside several of Oldenburg's gates for centuries, the first significant settlement outside the city's gates did not occur until after 1800 (see figure 2.3).[29] These developments were reflected in the creation of a new category of suburb (*Vorstadt*) for extramural settlements of urban character in the 1833 city code. In this transitional period, the nightly closure of the city gates remained a significant practical division separating Oldenburg from its external settlements.

Pressure to keep Oldenburg's gates open at night increased after 1833, since the inhabitants of those new suburbs gained the rights of city dwellers. The magistracy received a number of petitions for eased movement in and out of the city. In 1834 the inhabitants of the Heiligengeist Street north of the city petitioned for a special exemption from the payment of the nighttime gate tax, pointing out that as residents of a suburb they suffered under the

2.2 Entering the city from across the Hunte River outside the Damm Gate, where many Oldenburgers and laborers living outside the city traveled back and forth every day. *Vor dem Damm-Thor zu Oldenburg im Jahre 1820*, hand-colored on paper, Niedersächsisches Landesarchiv Oldenburg, NLA OL K-ZE Best. 298 OL A Nr. 1.

double burden of both contributing to the city funds for night watchmen and other urban services while having to pay a tax every time they entered the city after dark like other outsiders. The magistracy refused their petition, complaining that they were trying to demand for "suburbanites" (*Vorstädter*) special privileges not even accorded to townspeople.[30] As resentment of the gate closure increased, gate guards became the target of attacks by rowdy young journeymen. The gatekeepers were the objects of much derision, their duty mockingly referred to as "sleeping at the gates."[31]

As in Leipzig, it was not just practical annoyances that motivated the abolition of the gate tax but also anxiety about what the nightly closure of the gates said about Oldenburg's status as a modern city. The injustice of the gate tax became a favorite topic in the local press, which was beginning to flourish in Oldenburg in the 1840s. The argument launched against Oldenburg's status as a "closed" city combined reasons that were both symbolic and practical. In its first year of publication (1843), the liberal journal *Neue Blätter für Stadt und Land* (New pages for city and country) ran an article lamenting

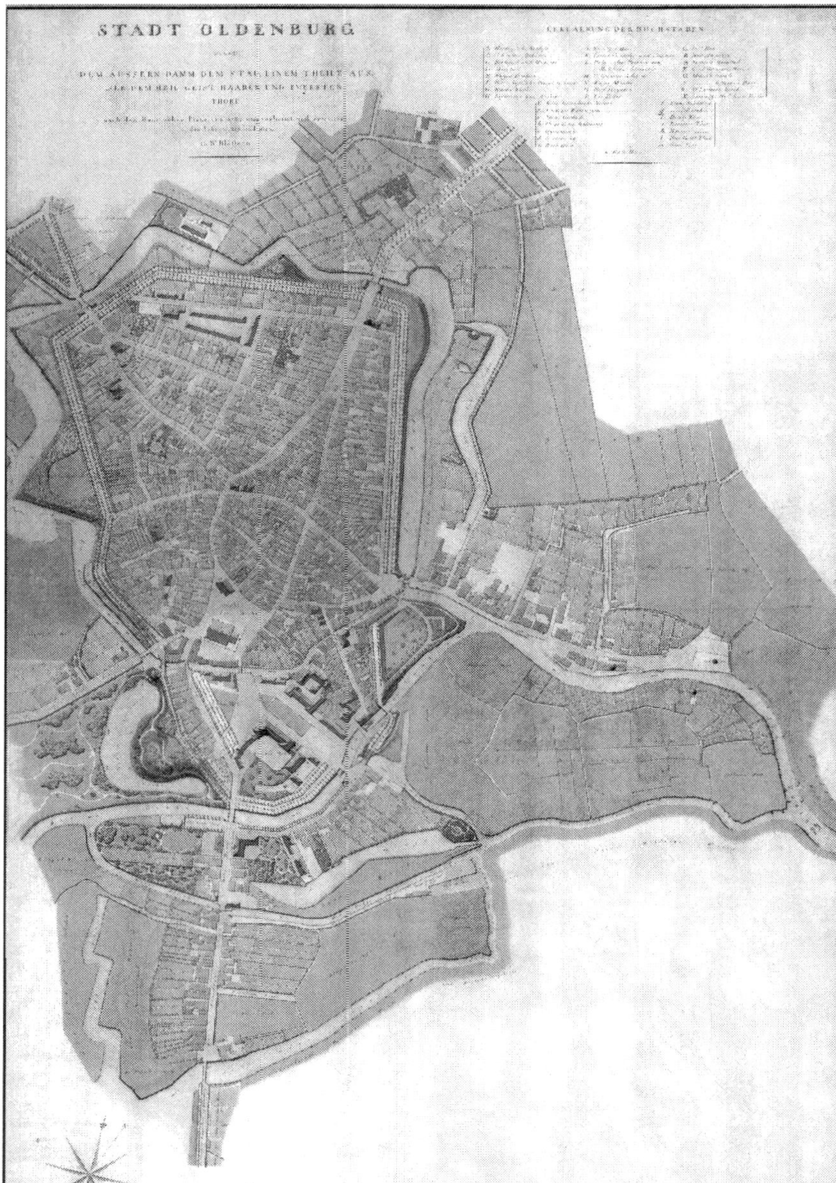

2.3 Oldenburg gained developments beyond the city's moats in the early nineteenth century, as seen here in the 1820s. H. A. Vollimhaus, *Stadt Oldenburg nebst dem aussern Damm, dem Stau, einem Theile ausser dem Heil. Geist, Haaren und Eversten Thore*, 1828, pen and watercolor on paper, Stadtmuseum Oldenburg, KP 495_498.

that Oldenburg's practice put it in poor company among German cities: "The custom, originating in ancient and dark days, of locking the gates in the evening and only allowing entry with the payment of a small fee, is still followed in all of Germany only in the northern Hanseatic cities, where much that is ossified and withered from age remains, and—in Oldenburg!"[32] Beyond the claim that barriers to free flowing traffic belonged to another, less enlightened time, there were three additional arguments made against the gate tax in the press and, eventually, by the city government itself. First, gate closure held up traffic without producing much in the way of profit. Since a number of families who rightfully belonged to the city had in recent years moved beyond the gates and "since so far no independent suburban life has been formed," it was only right that they should be able to move in and out of the city freely. Given the new suburban settlement, some of the gates even seemed to "stand in the middle of the city." Second, restricting traffic going in and out of the city, even at night, was unsafe and delayed medical help to the suburbs. Finally, as a communal tax, the gate toll was neither fair nor efficient.[33]

The final impetus for abolishing Oldenburg's gate closure, however, came as part of the preparations for a civic celebration. The city planned a large public commemoration for the five-hundredth anniversary of its city charter on January 6, 1845. The celebratory declaration printed for the anniversary made clear what it was about the city that its officials wished to commemorate: "We celebrate on January 6th not the half incidental settlement of some families of farmers and fishers at the juncture of the Hunte and the Haaren, nor the establishment of a castle by our sovereign prince on this spot . . . but instead the memorial day of the legal and documented foundation of the first city of our land, which signified in its time, the first refuge for free commerce and civic and intellectual development in the deep night of medieval barbarism."[34] What is notable about this formulation is the desire to deemphasize Oldenburg's founding as a product of its geography or an expression of princely power in favor of its founding as a refuge for urban freedoms. In marking Oldenburg's development as "a blooming and friendly capital," city officials sought to emphasize the contrast between the "tightly walled medieval city, fearfully clinging to the feet of the count's castle" and the open and modern city that Oldenburg had become in the nineteenth century.[35] Such old-fashioned customs as the gate closure threatened the clarity of this contrast. Seeking to persuade the public, the magistracy argued that since the gate tax originated in Oldenburg's long vanished character as a fortified city it was no longer a suitable method for the collection of city funds.[36] Social, economic, and practical considerations all factored in the decision, but ultimately the gates' perceived origins in the city's past as a fortress were used to bolster the case against their suitability in the present, even though there was no necessary link between the social and economic function of the city gates and the military function of the vanished ramparts and bastions of the city.

The abolition of the gate tax was greeted with general merriment. In a public demonstration, crowds of Oldenburgers cheered: "Hear it and believe it—long ago / There stood at the gate a dour man / night after night in the same spot— / Demanding money from everyone!" Even at the moment of its abolition, the practice of locking the city gates seemed a practice belonging to the distant past. Only thirty-five years after the gate closure ceased, the regional author Ludwig Strackerjan would write that the closure of Oldenburg's gates seemed "like a fairy tale from the olden days."[37]

Oldenburg's city officials fashioned a historical narrative of the city that equated the demilitarization of the court city with social and economic progress and urban growth. With this narrative they set the stage for the city's attempts to annex large areas of its hinterland in the following decade. But, although the discussion of the gate tax occurred entirely within the city community itself, Oldenburg's subsequent expansions required negotiations with residents of the hinterland who did not identify themselves with the historical city. These residents would challenge not just the city's attempts at annexation but also the understanding of modernization and progress as a process of transcending and erasing historical borders. In the debate over the city's subsequent expansion, some Oldenburgers would come to offer an alternative understanding of progress grounded in the unique environment of Oldenburg's hinterland and a particular understanding of environmental conquest. The middle decades of the nineteenth century were a period of reconsideration of what it meant to be a city and of the relationship between city and hinterland, but this process was not so simple as either the removal of the city ramparts or the opening of the city gates.

URBAN IMPROVEMENT IN THE ERA OF RECLAMATION

Since Oldenburg's changing borders were inscribed not just on maps and documents, but also in the very water, sand, and mud, the development of those borders cannot be understood in isolation from a larger story of environmental change. Nineteenth-century Oldenburgers understood the boundary between urban and rural land as one embedded in a landscape of boundaries between water and dry land, fertile farmland and barren heaths. In 1852 an author in the *Oldenburgische Volksfreund* (The Oldenburger people's friend) described both the land and the people of Oldenburg as divided by the sharp contrasts between geest, moor, and marsh. "Whether mountain or plain, heath or marshland, determines the nature not only of plants and animals, but also people," he wrote. The course of an individual Oldenburger's life was determined above all by whether he settled on the poor, barren land of the geest, the rich soil of the marsh, or the treacherous amphibious zone of the moors. In its flatness and monotony, he wrote, the geest resembled a desert or an ocean. Like those landscapes, the geest's high sandy expanses awoke in the viewer a feeling of "limitlessness" (*Unbegrenztheit*). The sea teemed with life, but here the usual fecundity of nature was nowhere to be found. A traveler might wander for days without seeing even a bird. Here, the author concluded, the arrival of the railway would be especially welcome, for it would allow the traveler to escape the landscape's monotony all the more quickly. The grazing for which geest landscapes were largely used led to erosion and impoverishment of the soil. Sheep, cows, and pigs foraged in the heath forests on the geest. Foraging opened the forest floor over time, and yearly cuttings to set by food for cattle in the winter further thinned the trees.[38]

By contrast, the marsh offered a wealth of resources and fecund soil and was populated by charming farms and the numerous cattle for which Oldenburg had long been known. But the fruits of the marsh were hard-won and hard defended because the marshland stood under constant threat of flood. A complex system of canals and channels, sluices and dikes (known as *Dämme*) allowed rainwater to drain away while protecting the land from rising sea or river water. These systems required regional coordination and oversight—led by an appointed director (*Deichgraf*). Inhabitants of the marsh had a saying that, without the dikes, they could afford "silver ploughs," but that dikes were the "golden rings" that created their charmed landscape. It was not all progress, though, for the more the land was drained for agriculture and settlement, the lower the level of the land and the more catastrophic the periodic floods became.[39]

Interspersed with the geest and the marsh were waterlogged moorlands. The moors of the region south of Jade Bay had formed as rising sea levels lifted the water table and saturated the ground, creating boggy land covered with moss and low plants like heather—but never any trees. The high moors to Oldenburg's west and south resulted when rainwater soaked the ground above the water table, keeping the soil nutrient poor and allowing the accumulation of layers of rich, spongy peat.[40] High moors foster little in the way of plant or animal life, but in this soil-like material the moors also sustained life. Blocks of peat—cut from the ground and dried for use as fuel—were a fixture of life not just in the moors but also in the cities of the German Northwest. In her memoir of growing up in late nineteenth-century Oldenburg, in a house with a garden along the northeastern edge of the inner city, Helene Lange recalls the mountain of peat, growing and shrinking with the seasons, that was a fixture of her childhood world.[41] Traditional cultivation in the moors involved the yearly controlled burning of the surface and planting of buckwheat in the warm ashes left behind. Living in the moor meant isolation. Roads were hard to build on the wet ground and were often passable only during the driest months. Up through the nineteenth century, "moorland was wasteland," a treacherous landscape that betrayed trust and resisted productive use or clear categorization as either dry land or wet.[42]

All three landscapes—geest, marsh, and moor—were transformed by nineteenth-century technologies and attitudes. Oldenburgers had been aware of the changeability of their environment, and the complicity of human communities in that change, long before the nineteenth century—as the geest grew with deforestation and diking transformed coastal marshes—but the industrial revolution and newly ambitious state-backed improvement projects introduced changes on a larger scale and at a faster pace. Life on the geest changed dramatically with the introduction of artificial fertilizers, and more ambitious canalization and water control systems affected both marsh and moor. Peat production became a large-scale industry, as new excavators and presses allowed the harvesting of larger quantities. The arrival of trains to Oldenburg, which in their early years also ran on peat, enabled its export.[43] Finally, Oldenburg launched itself (belatedly, compared to its Frisian neighbors) into opening up the moorlands through inner colonization programs in the second half of the nineteenth century, planning numerous settlements along newly dug canals.[44]

Common to these projects was a nineteenth-century belief in progress and the power of technology and entrepreneurship that changed attitudes toward agriculture and the barren marginal lowlands that surrounded most German towns and cities in the north German plains. From this new perspective, the traditional economy of extensive communal use of so-called borderland (*Grenzland*) or wasteland (*Ödland*) appeared primitive rather than resourceful. New attitudes toward water in particular transformed the "wastelands" of northwest Germany into "gold mines," waiting to be mined through improvement and regulation. In the eyes of planners and experts such as hydraulic engineers, modern principles of engineering and aesthetics conspired in the celebration of deepened and straightened or evenly bending waterways and a preference for quickly flowing water and the elimination of swamps as eyesores.[45]

These projects were driven by enterprising individuals such as Ludwig Starklof (1789–1850), a liberal bureaucrat and polymath who spoke of the power of canalization to transform not just environments but people too. Canals, he wrote, would not just connect Oldenburgers to the wider world but would also "wash" away the dirt and poverty of Oldenburg's rural areas. Life in the fen colonies developed along canals was supported by agriculture and the digging of peat. This was a linear model of settlement without centers—communities developed along canals, without town squares or clear borders. Life in the fen colonies was hard, as even an enthusiastic booster such as Starklof admitted. As with all artificial waterways, these projects were plagued by problems in maintaining water levels, necessitating the construction of feeder canals and reservoirs. The promise of colonization was not just improvement of the landscape but also the provision of land for the many already leaving for the Americas each year.[46] Envisioning a transformed rural Oldenburg that would rival the fen colonies of East Friesland, Starklof particularly hoped for a canal between the Hunte and Ems Rivers. Although his dream would be fulfilled eventually, Starklof himself never saw the canal construction begin. Delayed by the revolutionary upheavals of 1848–1849, work did not begin on the canal until 1855, and it took nearly forty years to complete. Even in the early stages of its planning, however, the idea of a new canal connecting the Ems and Weser Rivers changed Oldenburg's relationship to its hinterland.

The rural landscape of Oldenburg was transformed, and the status of the city of Oldenburg as a regional hub was also shaped by the changing relationship between the state of Oldenburg and the wider German world. In 1853 the Grand Duchy of Oldenburg joined the German Customs Union. The following year, Oldenburg and Prussia reached a historic agreement, which paved the way for a political bargain between Prussia and Oldenburg that allowed the foundation of a port city for a Prussian navy. This was the city of Wilhelmshaven, less than sixty kilometers to the north of Oldenburg's capital, on the western edge of the Jade Bay: a "town on command," or even a "city against its will," made possible only by a heroic struggle against nature.[47] Prussia purchased Wilhelmshaven from Oldenburg because the Jade Bay provided sheltered and ice-free access to the North Sea in a relatively isolated location. But it was anything but a natural position for a city. It took huge expenditure, thousands of laborers, and over two decades to transform the coastal marshes into a city. When Wilhelm visited the city to christen the completed harbor in 1869, a witness to the ceremony later recalled that the Prussian king had "looked out in every direction at a wasteland."[48] Only with the building craze of the 1870s would the city gain the familiar urban fixtures of government buildings, paved roads, town squares, and parks. Even after all these improvements, it remained hard to keep the water out. Problems with drainage, sewage, and sanitation continued to plague the city well into the twentieth century. Wilhelmshaven was the most ambitious project of urban development through land reclamation in the "amphibious zone" around the Jade Bay, but it reflected the larger regional projects to transform marsh and moor in the state of Oldenburg.[49]

These regional projects necessarily altered how Oldenburgers thought about the relationship between the city and the hinterland—and even about what a city could, and should, look like. In addition to changed regional power relationships and transportation networks, a shared aesthetics of rationalized and legible space connected rural and urban improvement plans. Engineers and urban planners applied a common set of principles of improvement, prioritizing symmetry, ease of movement, and clear boundaries between water and dry land.[50] One of these figures was the city planner Ernst Friedrich Otto Lasius (1797–1888). In 1853 Lasius published a pamphlet entitled *Blicke in der Stadt Oldenburg, Vergangenheit und Zukunft* (A brief look at the city of Oldenburg, past and future), which outlined his ambitious plans for Oldenburg's future development, linking the capital's expansion to regional transportation and reclamation plans (see figure 2.4). Much as Starklof envisioned transforming both landscapes and people with canals, Lasius envisioned transforming the city by making its landscape legible and symmetrical and by rationalizing the relationship between city and natural environment.

Lasius had been deeply involved with the last stages of the city's defortification. He served on the commission in charge of the removal and redesign of the city's western ramparts in the 1840s. In the 1830s the first major construction project outside the gates had created the horse market (*Pferdemarkt*), hospital, and barracks building, which today houses a library to the north of the city, but Lasius was concerned that the city develop symmetrically, not just expanding into the geest to the north. In the following years Lasius dedicated himself to planned expansions of the city in marshy land to the west and east as well. Under his direction, the city gained a new harbor, as the Hunte was straightened for increased shipping traffic. A grand central post office was built on the former fortifications land in the east, near the new harbor, in 1846.[51] A *Fluchtliniengesetz* (law on street lines) provided guidelines for future street construction in 1847. Although his plans were characterized by a certain shortsightedness (he planned to place the train station all the way across town from the city's busy port, for example), Lasius thought about and planned for Oldenburg's future development in a new way.

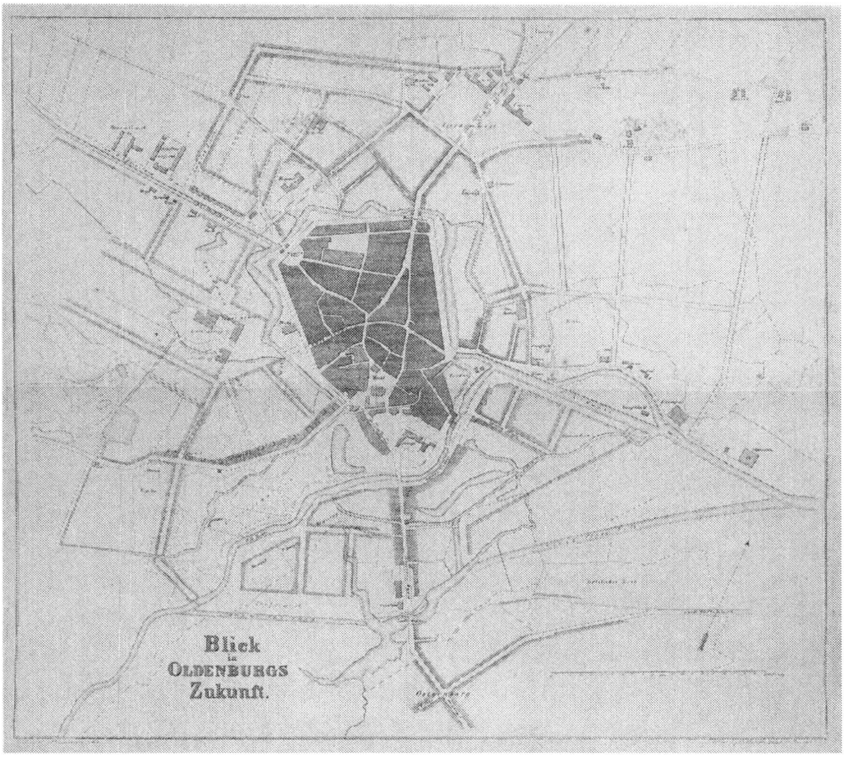

2.4 The plan Otto Lasius included in his pamphlet on the future development of Oldenburg, showing how he hoped to "round out" the development in the marshy areas to the west and east of the city. Lasius, *Blicke in der Stadt Oldenburg*, foldout map.

He strongly objected to what he saw as the "absentmindedness" (*Zerfahrenheit*) that characterized new building at the city edge after the removal of the fortifications. He sought more orderly, planned developments. He wished to fill in "holes" of rural character in the urban edge and looked forward to "the prospect of a rounding off [*Abrundung*] of the city's expansions," even to the west, where growth had been most restricted by the swampy land of the Dobben. He wished above all to defend the city against the "dispersal of sporadic additions."[52] Much as moorland became spatially rationalized and productive with straight-line canals, Lasius believed that the city became a productive space when organized and geometric.

Lasius reflected on the relative advantages of planning new developments on high, dry ground, where construction would be easier, and pursuing more complex and costly reclamation projects to allow more symmetrical developments all around the city. One advantage of pursuing developments in the boggy ground to the south and west would be that the necessary draining of the surrounding land would "distance the quantity of stagnant water more and more from the city, and so win an improved state of health" for the entire settlement. He saw Oldenburg's urban expansion as linked to regional development projects. He looked forward to the introduction of railroad lines and canals to Oldenburg, which would allow cheap and efficient transportation of sand to build up the marshy ground, so much so that he projected construction in the Dobben marshes would be hardly more expensive than building on dry ground. He planned the theater district (the Theaterwall) along the western fortifications land, and an expansive new residential neighborhood in the Dobben near the palace gardens. After 1860 the architect Hero Diedrich Hillerns took up Lasius's idea for an expansion to the west. Hillerns and Lasius together created what would become Cäcilienplatz, and they planned a checkerboard of grand residential streets west of the new plaza.[53]

Beginning in the middle of the nineteenth century, urban expansion planning was inseparable from environmental improvement projects, both in the city's immediate hinterland and as they transformed the wider region. Both urban reformers and fenland colony boosters were motivated by the desire to find new land and opportunities for development at a time of rising overseas migration.[54] In Oldenburg's environment of moor, marsh, and geest, urban borders and urban expansion plans were embedded within regional projects of moorland reclamation that reflected centuries-old struggles to reshape the land for stable settlement. As the construction and removal of Oldenburg's extensive fortifications had made evident, changes to the urban borders implicated water control systems linking the city to its hinterland. As the city grew beyond its early modern borders, Oldenburgers applied the framework of environmental conquest and crisis to understand the changing relationship between the capital and its hinterland. In discussing the proper limits of urban growth, a vocabulary of water, sand, flood, and reclamation linked local questions of class and political power to regional, national, and global narratives of modernization and the conquest of nature. From this perspective, borders did not denote the past but, rather, a landscape of progress and environmental control.

GROWTH AS FLOOD AND CONQUEST IN OLDENBURG'S HINTERLAND

In the decade after the opening of its gates, Oldenburg's city government set its sights on a vast area for expansion, presaging Mayor Goerlitz's urban imperial ambitions in the next century. At this moment, just after the elimination of the gate tax and as the city was first considering borders drawn without any reference to the old fortifications land, administrators showed a remarkable flexibility in imagining the future of the city. The city council even considered making urban citizenship independent of residence.[55] These new territorial ambitions were made possible by the environmental conquest of the city's hinterland. The mid-century debate over the appropriate limits for Oldenburg's expansion conflated the vocabularies of each environmental and political control: the growing city both a flood and a conquest.

At least three perspectives on the advisability and proper limits of Oldenburg's urban expansion emerged in the ensuing negotiations: that of the city government, that of the residents of the city's mostly rural hinterland, and that of the state government. Urban administrators argued, on behalf of the city's interests, that unencumbered growth was the right of any city, and of a capital city in particular. They argued that Oldenburg's urban growth was the natural result of its function as the state's central place, the seat of the state bureaucracy, and a military garrison.[56] In the language of a report submitted to the grand ducal government by the city magistracy in September 1847, new developments at the city edge were nothing more than the "natural expansions of the city, beams that radiate from the city as their center." The report went on to compare the suburbs to the "limbs" of the city "body."[57]

Oldenburg County, the administrative unit responsible for the area surrounding the city and including communities of Osternburg, Holle, and Wardenburg, argued against the city's proposed expansion. Although the city attempted to equate the public interest of the state with the best interest of its capital, from the outside it seemed that the city was trying to further its own interests by expanding at the expense of the general interests of the state. Suspicious of the city's attempt to collapse the distinction between capital and state, representatives of the county argued that the capital city could not claim exclusive credit for its rapid growth. The city may have been the center of commerce and government, but it "instead had the entire state . . . to thank" for its prosperity and growth. They countered the body/limb metaphor of the magistracy with their own, writing that "the capital is the heart from which the most varied blessings flow through the arteries of the whole land. No one would allege that the city derives from this state of affairs the right to call on the entire land to share its burdens."[58]

As city administrators made the case for urban expansion, the state government cast itself as mediator between the city and its hinterland, affirming the equality of the urban and rural communities and requesting that residents of the areas to be added to the city be consulted on the question of their own incorporation. Residents of the city's hinterland in the area of the city's proposed expansion and their representatives differed in their reactions to the prospect of becoming city dwellers, depending on how much their own lives centered on urban business and culture and how able they were to support urban taxes. Many, however, found the prospect unappealing. Belonging to the city appeared to be a bad economic deal. The financial burdens associated with town living consisted of a military billeting tax (*Servicelast*), taxes to fund urban services such as night watchmen and street maintenance, the tax on fuel and cattle (*Octroi*), and the poor tax.[59]

The relationship between the state of Oldenburg and its growing capital city provoked a spirited exchange in the pages of the *Neue Blätter für Stadt und Land*, in which practical and functional arguments about finances and administration confronted abstract concerns about the formal relationship between the city and the state. An article published in August 1847 accused the city of cloaking its financial interests and questioned the city magistracy's claim that it wished to annex only those areas with an "urban appearance," when in fact houses within and without the city did not look any different, and those who lived outside the city had consciously chosen a rural life by electing to settle outside of city limits. The same author concluded that the fate of the city district formed by Oldenburg's most immediate hinterland was sealed. Without economic independence, it would never be able to defend itself from the incursion of the city.[60] An advocate of city expansion writing in response claimed that the city had an inherent right to growth unlimited by anything but the borders of the state: "In my opinion, every city has acquired through its very founding the right to expand all the way to the state border [*Grenze*]: because stagnation is unthinkable, a city without the right to grow would be destined from the start to become smaller again." The author pointed out how ridiculous it was to restrict a city to its historical borders by comparing the city of Oldenburg to other capitals—Paris and Berlin—that had many times expanded their limits beyond old walls and fortifications. It was unreasonable to expect a city to interrupt its "natural development" in order to preserve the peace of a few inhabitants of its environs. In other words, the details of the situation on the ground were insignificant when compared with the necessity of Oldenburg's growth as determined by its very nature as a capital city.[61]

Another step toward the erasure of the old fortifications line as a practical urban boundary was taken in 1849, when the city proposed eliminating the legal distinctions between city and suburb only sixteen years after the category of suburb had been introduced.[62] With this, the old fortifications line ceased to be a stable reference point for the historical city, against which its growth could be measured. Some took this as a sign that the very notion of the city, by definition a controlled and contained settlement, was becoming obsolete. The president of the ducal cabinet, Johann Friedrich Mutzenbecher, speculated that "in the legal respect there will in the future be no more cities, and the population of those who do not primarily engage in agriculture, which one until now would have called *city folk* [*Städter*], will, as is already happening, spread themselves more and more throughout the land." Although Oldenburg retained an overwhelmingly agrarian character in the mid-nineteenth century, fenland colonization programs also changed Oldenburg's seclusion and status as a central place.[63] Instead of being surrounded by empty and unproductive land, Oldenburg's wider hinterland was now a space for future development and improvement. The city government assumed that the best interests of the city were synonymous with the public good, and those outside the city could appeal only to private, individual interests. The state, on the other hand, interpreted city and hinterland as equal partners in the negotiation over incorporation.

A series of meetings with residents outside the northern gates in 1852 failed to secure support for the city's desired expansions.[64] A committee representing the city territory (*Stadtgebiet*)—the rural area inside the city district but not of sufficiently urban character to have become a suburb in 1833—challenged the rationale for the city's proposed expansion. They contested the city magistracy's claim that new developments in the hinterland had occurred only because of money and residents flowing from the capital and that they should be taxed as part of the administrative city: "The causes [of growth] are to be found in the development of bourgeois society and above all in the increased development of the inherent powers of capital, of which more and more is drawn originally from the land through wealth gained by agriculture. It would therefore be a completely unjustified and incorrect view to attribute the growth of the population in the city territory to emigration of city residents over the city borders, as the magistracy seems to do." Instead of joining the new developments with the city, the committee proposed that sections of the city should be given over to the surrounding district. It seemed perverse and an "injustice against present and future," they argued, to respond to the "blooming" of new communities outside the city borders by chopping them up and depriving them of their sources of income.[65] In response to these challenges, the city government increased its ambitions, rather than scaling them back. City administrators began considering a more extensive urban expansion, looking not just to the new developments just outside the city but also to a wider area including much land that was as yet undeveloped, including significant portions of Eversten, Osternburg, and Donnerschwee.

This next stage of the struggle over the planned urban expansion increasingly became a showdown between the city of Oldenburg and the working-class center of Osternburg, a small but rapidly growing community southeast of Oldenburg on the other side of the Hunte River. Early residents of Osternburg, thinly settled before the nineteenth century, eked out a hard existence as farmers on the sandy deposits in the Hunte valley. Like many residents of Oldenburg's hinterland, they supplemented their relatively meager agricultural endeavors by gathering peat from the surrounding moorlands. Around 1800 the parts of Osternburg closest to the city were settled by a combination of court and state employees and craftsmen, especially along the road leading out of the Damm Gate (known as the "Damm"), over the Hunte River and one of its branches (called the Oeljestrich) to the center of Osternburg.[66]

Osternburg's growth in the nineteenth century was made possible by new projects of environmental conquest, both moorland reclamation projects nearby and also distant diking on the North Sea coast. If it were not for the diking measures taken to protect the coastal settlements, water would eventually have progressed to the edge of the geest and Osternburg today would be part of the coastal marsh region. Neighboring farmers in Holle and Blankenburg had to fight for their fields against periodic floods with diking on the Hunte and diking and dams on the edge of the geest, a problem progressively worsened in the nineteenth century by the straightening of the Hunte River. As throughout the region, the same measures that made settlement over a wider area possible actually worsened the periodic calamities of flood. In the 1840s Osternburg was surrounded by moors that had long been uninhabited, but which were now newly settled by a scattering of moorland colonists. According to an 1844 report, ninety people lived in the Drielakermoor to the northeast, most of them new colonists. In Tweelbäke there were also a number of new colonists' settlements.[67] These moorland colonists lived off agriculture and harvesting peat from the bogs. A diverse settlement throughout the nineteenth century, Osternburg's population remained divided into three distinct groups with different interests. In addition to the moor colonists, there were farmers who lived in Bümmerstede, Neuenwege, and in parts of Osternburg proper, and craftsmen and high officials who lived on the Damm Road leading into the city.

Osternburg's character really began to change in the middle of the nineteenth century. First a glassworks was established in 1845, and later, the digging of the Hunte-Ems Canal (begun in 1855) and the construction of a large new spinning mill in 1856 transformed Osternburg into a small industrial center. As it grew, Osternburg remained a spatially divided community with two distinct centers: one for middle-class residents and state officials oriented toward Oldenburg and the street outside of Damm Gate that had been among the capital's first expansions, and a second for workers, which maintained a more self-sufficient culture and community in Osternburg.[68] The first center, nearer to the Damm city gate, was characterized by single-family houses, stately villas, and high tax revenue producing households. Osternburg's working-class center was more densely settled, characterized by employer-owned worker housing, like that built with the spinning mill in 1856. These workers came not from the city itself, but from throughout the region, and especially East Friesland. By 1887, Osternburg would have six thousand inhabitants. Many of the new arrivals were factory workers, including a large proportion of fluctuating migrant workers.[69] These changes were just beginning to get under way during the 1850s, when Oldenburg and Osternburg first clashed over the possibility of annexation. At this point Osternburg was still primarily a rural district. While later tension between them in the 1890s through the 1920s concerned the suburb's industrial character and the presence of large numbers of workers, the earlier conflict centered much more on the nature of agriculture, reclamation, and the environment.

On March 17, 1854, the collected wardens (*Hausväter*) of the districts affected by the proposed expansion were brought together to consider the plan and vote their support or rejection. Of the 197 wardens invited from the outer Damm, Osternburg, Wunderburg, and Drielaker Moor 113 came, representing each moor, agricultural, and city-oriented settlement. With the exception of three craftsmen, all of those present rejected the planned expansion of the city. After the meeting, 32 of the wardens from Osternburg produced a report explaining their reasons.[70] They acknowledged that "certainly the city could no longer find room for the development which it had experienced over the last thirty years within its ramparts and moats" and that expansion beyond this "unnatural confinement" was inevitable. They found the extent of the planned expansion, however, ridiculous and unnatural, all the more so since large parts of the area planned for inclusion within the new city borders were pasture lands that "stood under water for the greater part of the year," giving Oldenburg an appearance "as if it lay in a lake."[71]

The accusation that the city risked confusing the boundaries between dry land and wet land was a complaint that had particular resonance in Oldenburg where the construction and maintenance of clear boundaries between water and land through water control systems was crucial to the area's livelihood. The image of the borderless city was that of a flood: the city would appear as if surrounded by a lake. The city of Oldenburg had based its claim to this region on the urban character of either its land or its inhabitants, but to Osternburg's representatives this claim seemed in direct denial of the situation on the ground. They represented this trespass by appealing to one of the most sacred—but also most tenuous and closely regulated—boundaries to be found in the local landscape of marsh and moor. The Osternburg representatives compared the difference between city and land to that between geest and marsh: "Just as it would be completely impossible for an infertile geest district to bear the costs of the dikes and sluices necessary in marsh regions, a rural population cannot bear the costs of urban institutions and establishments."[72] Where elsewhere the elimination of urban boundaries evoked progress and modernity, to the Osternburg representatives eliminating boundaries meant going backward: relinquishing control over the landscape. Notably, neither the city nor the state government used this kind of reasoning. The explanations they provided for the proposed expanded boundaries made little mention of the natural landscape, speaking instead in legal and economic terms.

Although the financial burden associated with city dwelling was a primary concern, the Osternburgers' protest against the city's proposal framed the question in much grander terms than financial self-interest. The Osternburgers characterized the city magistracy's attempt to incorporate the countryside into the legal city as an "Asiatic" strategy more suited to the nomadic peoples of the Mongolian steppe. Their explanation is worth quoting at length:

> Throughout Europe, wherever a village is near a city, the influence of this proximity makes itself felt just as in Osternburg, but only in the Asiatic steppe does it happen that urban settlements, founded by peoples who are really of a nomadic lifestyle, encompass within themselves districts, such as that from Drielake to the toll barriers on Vechtaer Road, where the farmer fertilizes and plows his land, cuts his hay and pastures his cattle. . . . In Asia it might be appropriate, where nomads have been gradually accustomed to fixed settlement, . . . to give inhabitants the opportunity to carry on their earlier modes of life within the so-called city,—and according to what one reads in travel descriptions, the Mongolian cities may well have naturally grown quarters of a kind similar to those that the city of Oldenburg would like to incorporate.[73]

The use of an imagery of the steppe and Mongolian nomads is more than an arbitrary attempt to evoke the foreign, "primitive," or unfamiliar. The steppe environment—high sandy grassland—mirrored the ways in which the new settlement area had been won into Oldenburg's suburbs, by carting in loads of sand to raise the land. In their desolation, the moors had long been described as primitive steppe environments. See, for example, the often quoted descriptions of the travel writer Johann Gottfried Hoche: "One believes oneself to be in the steppes of Siberia when one wades through the moors. . . . Everything is desolate [*öde*] and still. . . . Not a tree, not a bush offers shade."[74] The Osternburg commission accused the city of essentially undoing the progress achieved in recent years. Moorland reclamation projects made progress by establishing clear boundaries between dry land and wet where once everything had been an amphibious zone. Moreover, the vocabulary used by the Osternburg representatives to describe what was at stake in Oldenburg's expansion plans evoked larger ecological processes. In the discussion of Oldenburg's growth, sand, steppe, and desert symbolized borderlessness and growth without limits.

Osternburgers found Oldenburg's planned expansion especially threatening because it had no clear boundary or fixed border, no wall or demarcated edge. Not only did the city threaten to swallow what was clearly without urban character; city officials also refused to draw a line beyond which they would not progress. All of the grand duchy seemed an open frontier for its ambitious capital city. "There is one point that distinguishes these Mongolian cities from what the city of Oldenburg is proposing," the Osternburg representatives' letter continues: "they surround their entire territory, including field and pasture, within a wall, and with this achieve a fixed border between what is inside and what is outside. This fixed and unmistakably marked border is of great importance to the essence of a city, and everywhere where history tells of city founders we find that they began their work with a moat or a wall to decisively separate city and country. Conversely, where cities in recent times draw idealized lines in the place of old customary divisions they . . . bring about nonsensical relationships." The Osternburg representatives pointed instead to limited construction projects just beyond the old city's borders as more appropriate, modest expansions. The city had only very recently abandoned the project of physically demarcating the urban edge—with hedges, fences, ditches, and rebuilt gates.[75]

Although acknowledging that the only practical use of "rampart and ring wall" was to lend a "palpable border" (*greifbare Grenze*) to the extent of urban services, the Osternburg representatives argued that this reduced function remained significant even in an era of ongoing urban growth. They saw Oldenburg's planned unwalled incorporations as an attempt to expand the circle of the city's ability to extract profit far beyond the practical reach of urban services and prosperity into the relatively impoverished hinterland, since the expanded city proposed could not reasonably be reached by urban street lighting, plumbing, fire protection, or water control for "centuries" to come.[76] In their minds, this confusion of borders was equivalent to denying the differences in kind between the rich land of the marsh and the poor soil of geest and moor that so shaped life in Oldenburg.

Faced with this resistance, Oldenburg did not succeed in its ambitious plan to incorporate Osternburg. When the new city borders were finally set in 1856, Osternburg retained its independence, though it did lose a significant portion of its wealthiest territory. The Outer Damm Road, connecting Oldenburg to Osternburg, became part of the city. Osternburg's citizens demanded reparations for the lost territory. In the words of the Osternburg council: "What had belonged together for centuries was torn asunder."[77]

In the 1850s Oldenburgers disagreed over the significance of visible boundaries marking the border between city and countryside. In the following decades Oldenburg and Osternburg continued to struggle over the possibility of further expansions. But in the meantime, the context for thinking about the significance of urban borders also shifted. After German unification in 1871, the emergence of a landscape conservation movement and a new approach to the landscapes of moor and marsh meant that historical borders took on new meaning as visible markers of the connections between community and environment. In the late nineteenth century, Oldenburg's fortifications came to represent the particular logic and local peculiarities of the city's historical founding and development, both topographical and strategic. Exceeding the historical borders of the city was both a fulfillment of Oldenburg's destiny as a well-chosen, strategically located settlement and the overcoming of what made it unique by the universal logic of urban growth. In debates over Oldenburg's city border, local actors appealed to the universal characteristics of the modern city to frame particular interests and local circumstances.[78] It was not, in other words, just the border between the urban and rural that was contested, but the border between the universal and the particular, the general and the local. Urban planners and advocates of urban growth pointed to Berlin and Paris as models for Oldenburg's own path of development—they were, after all, each capital cities—but to opponents this looked like comically outsized ambition. Identity as a modern city might promise the right to growth, but not every mayor could become an urban Alexander the Great.

THE URBAN BORDER AS LANDSCAPE AFTER 1871

At the beginning of the nineteenth century, moorland was wasteland, but by the end of the century, Germans no longer equated wasteland with untouched and unproductive wilderness. This shift in how the landscape was perceived also shaped urban planning and the relationship between the growing city and its hinterland. As industrialization transformed natural landscapes not just through their improvement but also more evidently through their destruction, "waste" came to denote landscapes of human devastation as well as the un-useful (and classically un-beautiful) land of the moors, the steppe, and the desert. This changed understanding of what constituted wasteland was the shadow side of the nineteenth century's new understanding of wilderness as unspoiled nature, an environment to be protected from the ravages of industrial and capitalist exploitation. The transformation of the land through increasingly intensive utilization of natural resources gave rise to the first landscape preservation movements in Germany in the 1880s, reflecting a new appreciation for each nature and homeland (Heimat). While the Heimat movement allowed Germans to appreciate national identity through their local sense of rootedness in home communities, Thomas Lekan has further argued that "landscape preservationists fortified the ability of Heimat to bridge the local and the national by envisioning German strength emanating upward from the country's mosaic of unique regional landscapes, rather than downward from the Reich capital of Berlin."[79]

Historical German cities became objects for preservation during the same time that a new landscape preservation movement gave greater attention to the natural landmarks of their hinterlands, and the two movements were closely linked by common motives and, in some cases, personalities. In Oldenburg, this entailed a new appreciation for the beauty of moor and geest as more than neglected wastes at a time when reclamation projects, artificial fertilizers, and mechanical harvesting of peat threatened the survival of those landscapes in new ways. Painters such as Richard Tom Dieck and Otto Modersohn discovered the beauty of the moors as a landscape, mirroring the aesthetic reevaluation of the fens in English painting of the nineteenth century. Through the work of the nearby Worpswede artists' colony north of Bremen, paintings of the moorlands of the German Northwest played an important role for the break with historicism in German painting. Moorland and heath landscapes became popular subjects for both Heimat painters in Oldenburg and the visitors from Berlin and Dresden who settled in Worpswerde, but their approach to these landscapes was distinct. Local artists who had grown up in Oldenburg, in close proximity to the moors, saw these landscapes as an expression of belonging, instead of as a refuge from the homelessness of the big city.[80]

In the same decades as this aesthetic reevaluation of the moors, the Heimat movement in Oldenburg found new value in the culture and customs of human life in the moor. Earlier cultural evaluations of those who settled in Oldenburg's moors had rejected them as backward, impoverished, and in need of improvement and modernization.[81] An 1867 article in the popular illustrated magazine *Die Gartenlaube* (The garden arbor) labeled the moors of Oldenburg and Hanover as Germany's "Muffrika." Challenging this older narrative, Heimat historians and artists around the turn of the century instead celebrated moorland life as distinctive and culturally rich. The best-known example of this can be seen in the paintings of Bernhard Winter, which lovingly represented a romanticized view of everyday life in rural Oldenburg, such as the harvesting of flax.[82]

The reevaluation of the landscape—both environmental and cultural—of moor, marsh, and geest as something beautiful and worthy of preservation changed how Oldenburgers approached the place of the city within the regional landscape, as well as shifting their understanding of the place of Oldenburg's local landscape within German national space. On the side of regional planning and hydrography, the early years of the twentieth century saw the first acknowledgments that the moorland might be a landscape worth preserving. On the side of city planning, this entailed a challenge to the classicist city planning that had dominated Oldenburg's architecture and urban design since the early part of the century through figures like Adolph Stahr, Otto Lasius, and Hero Diedrich Hillerns, and which continued to shape the Gründerzeit (founders' era) suburbs of the 1870s. This planning tradition was confronted with a new longing for forms of architecture and urban planning that reflected Oldenburg's regional character and distinctive natural landscape. One element of this impulse was nostalgia for what had been lost in Oldenburg's modernization, in particular fond reminiscences of the largely disappeared walls and ramparts that had at one point so clearly marked the relationship between the city and the landscape in which it was embedded. Late nineteenth-century figures like Oldenburg's garden director Heinrich Ohrt lamented the loss of Oldenburg's peculiar natural heritage—its landscape—to the generic streets of its suburbs. Ohrt recalled how the walls of the inner city had earlier been a place from which one could view the distinctive landscape of Oldenburg's hinterland, especially the Dobben marshes, but now all one could see where once there had been marsh were typical checkerboard residential suburbs. In this, Ohrt typified a mode of thinking that was common in the late nineteenth- and early twentieth-century Heimat movement, which advocated the preservation of local landscapes as an alternative to the homelessness of the big cities.[83]

Identifying Oldenburg as a "natural" city—meaning a settlement that had developed spontaneously over centuries—became more important in the context of regional improvement projects. In 1888, the regional author Franz Poppe opened his account of the city's history by observing that that there was nothing coincidental about Oldenburg's location and history: it was no "accidental city" (*Zufallsstadt*) whose formation resulted from a "political bargain."[84] Instead, the capital city's development reflected a natural environmental logic in its location. Although Poppe contrasted Oldenburg with other small German capitals, his observation meant something very particular in a region where the creation of new cities and new land from water was an active and ongoing state project. He was likely thinking of Wilhelmshaven, the city by command formed by political bargain just sixty kilometers to the north. The Prussian port city was not the only city by command in the German Northwest either. Bremerhaven and Geestemünde, at the mouth of the Weser, were also port towns established in the nineteenth century as competing centers for trade and emigration. More generally, Poppe wrote at a time when these new ports were just one piece of a landscape of newly founded settlements in the moorland colonies. Of course, the boundary between the natural and artificial city was not such a clear one as a Heimat author such as Poppe might have wanted to make it seem. Oldenburg's development was dependent on the draining and reclamation projects just as was the "accidental" settlement of Wilhelmshaven.

A consequence of this new interest in the environmental logic of Oldenburg's original settlement was a romanticization and naturalization of the city borders, ramparts, and walls themselves. Perceived as part of the city's history as an organic settlement, the old walls and moats became like part of the environment itself, an expression of the link between the city and its landscape. Karl Jaspers in his memoirs of growing up in Oldenburg, which he wrote just before 1900, provides one example of this tendency to romanticize the past and to treat Oldenburg's extramural growth as a departure from the natural logic of the relationship between city and landscape. Jaspers contrasts the organic city in the marsh with the inorganic deserts of the newer suburbs beyond the old moats and describes the shift away from the natural logic of the old bordered city:

> As I grew up, I studied Oldenburg's old fortifications, the position of the castle, and experienced why one would settle here in the marsh, because here two tongues of sandy land, one from Bremen and one from East Frisia meet. . . . Then I sensed the dreariness [Öde] of the modern, straight streets outside the ramparts with their narrow front gardens and small back gardens, which were artificial plantings, neither natural nor garden art, but rather forced arrangements, made as if for puppets. . . . There was in the architectural style of Olden-

> burg after the war of 1870 a mood that, though I did not know why, oppressed and made desolate. The mood of grey rain, foggy windowpanes, ill-smelling garbage heaps and deserted construction lots characterized this city.[85]

In Jasper's account, the natural logic of the landscape of settlement represented by the premodern city has been disrupted by the imposition of a generic model of urbanism with its straightened arteries and equally spaced homes, characteristic of modern German cities but at odds with the city's local conditions and heritage. As the bleakness of the suburban streets evoked the bleakness of the barren landscape of the surrounding moors, the long vanished fortifications represented the naturalness of the premodern (and prenational) city.[86]

For Jaspers, the land outside the walls was a space divorced from nature: of puppets rather than people and artificial plantings rather than natural landscapes. Other memoirs of the era, like Helene Lange's, hit similar notes, lamenting the loss of Oldenburg's distinctiveness and connection to nature after 1871.[87] Jaspers uses the term *öde* (desolation or dreariness)—associated with the uncultivated land (*Ödland*) of the moors—to denote the suburb. Just like the moors, what made the suburbs desolate in Jaspers's description is their resistance to easy characterization: they were a "foggy" space of refuse and abandoned projects for improvement. In Jaspers's account, the city's old fortifications remained the border between the city and the wasteland. Although the concept of what constituted "waste" had changed—from the unimproved moors to the monotonous streets of Gründerzeit suburbs—it was still "a landscape that resist[ed] notions of proper or appropriate use."[88]

The conflation of natural landscape and the old city fortifications reached a new extreme in the writing of the regional Heimat author August Hinrichs (1879–1956) in 1935. In Oldenburg, he wrote, "marsh, moor, and geest" had been "for centuries an impenetrable protective belt [*Schutzgürtel*] separating the city from the outside world." But today, he continued, that same landscape had become the city's "living space" (*Lebensraum*): "Marsh, moor, and geest—and between them the city. But no longer divided, instead connected to each other through a thousand living forces."[89] In Hinrichs' account, the city's environment had gone from wall to landscape, from limiting boundary to Lebensraum, offering an understanding of the city's development that connected local place and growth.

By the early twentieth century, two rival accounts of the history of Oldenburg's growth had emerged. One account figured the city's nineteenth-century expansion as the expression of a universal urbanism freed from the restrictions of landscape. Building on Oldenburg's tradition of classicist city planning and continued by the urban imperialist Theodor Goerlitz, this was a version of ongoing expansion in which urban growth also meant overcoming the internal divisions of region and politics in the German nation. The second, represented by Heinrich Ohrt and Karl Jaspers, among others, romanticized that which appeared natural and unique to Oldenburg's historical and regional identity, interpreting the city's nineteenth-century expansions as a refutation of that old identity.[90]

By incorporating the distinct perspectives of the hinterland and environment in the story of urban growth, a dramatically different interpretation of the significance of the city border in German history emerges, one almost entirely excluded from existing narratives. In Oldenburg, city walls and ramparts signified more than political power and the corporate identity of the city. They also stood in for the tug and pull between human settlement and the forces of nature—the ways in which human beings had to reshape the land to their own uses. Premodern urban borders had given physical shape to long developed systems of water control and settlement. Nineteenth-century attempts both to extend control and to erase clear topographical evidence of the negotiated relationship between human settlement and environment were resented by those whose livelihoods in the urban hinterland most directly depended on systems of hard-won control of land and water. The late century rediscovery of the older landscape of fortification was part of an embedded Heimat identity that, in Lekan's formulation, was "between nature and civilization."[91]

The long history of conflict over Oldenburg's borders helps explain what was at stake in 1924 when a state legislator angrily accused Oldenburg's mayor of urban imperialism. The ambitious proposal for Oldenburg's expansion to which he was responding followed just a couple years after the incorporation of Osternburg as part of the city in 1922, finally achieving the ambitions of seventy years prior. This was an era of grand urban designs. The 1920 creation of "Great Berlin" had more than doubled the population of the republic's capital and largest metropolis, and all across Germany cities planned expansions. But the state legislator had some justification in accusing Mayor Goerlitz of outsized ambition. If the entire planned area had been incorporated, the city of Oldenburg would have become the fourth largest by area in the German nation. Here, in a lesser city center, the longtime ducal seat was adjusting to life without its grand duke following his abdication. Soldiers returning in 1919 generated a housing crisis, and the small city felt cramped as never before. In the turmoil of the 1920s, the city government's attempt under the direction of Mayor Goerlitz to expand Oldenburg's municipal boundaries was part of a project to present the city as dynamic and growing in the midst of economic troubles.[92] It also meant recasting the relationship between Oldenburg and Germany as a whole, seeing Oldenburg primarily in terms of its connections to the rest of the German nation. Before the war, Oldenburg had been a "garden city" and an "idyll." Its prior isolation meant that the threefold shock of war, revolution, and inflation had hit it even harder than the rest of the country. But in the wake of those shocks, Goerlitz hoped that even though it lay "in the furthest northwest," the city would become "one of the strong arrows the German fatherland required."[93]

Goerlitz clearly confronted the city of Oldenburg as an outsider. He identified as an expert urban administrator and technocrat who saw the city's problems in national and universal rather than in local or regional terms.[94] As the Worpswede painters saw the endlessness of the moor landscapes as a placeless antidote to the big city, rather than a Heimat landscape linked to the small city, Goerlitz sought in Oldenburg a universal model of urban growth in an era when professionalization of urban administration linked the city to the state in new ways, rather than as an expression of local identity rooted in history and landscape. Goerlitz considered the planned expansion a prerequisite for the city's further economic and political development: in order to make Oldenburg seem like a growing city, it was first necessary to grow the city. He further identified Oldenburg's artificially constrained borders as an example of a peculiarly German condition, where (in comparison to France) cities were small and oppressed by local nobility. The incorporations of 1922 and 1924, he wrote, righted an old wrong. Goerlitz associated the past with the local—and local divergence from a perceived norm of urban development as a misdevelopment in need of correction. The city's government had been attempting to annex Osternburg since the 1840s, but "petty interests" had "clouded the ability to perceive the importance of a future development."[95] By throwing out the urban frontier, Goerlitz sought to redress the balance between Oldenburg's past identity and its future development, finally securing its identity as a modern city.

Throughout the city's history, the fortifications land, both as defense and then as a space of display and recreation, served as a symbol of the power of Oldenburg's dukes and the relationship between the duchy and its capital. The diminishment of this border and the reclamation of land beyond unsettled this relationship and indicated a shift of power from the state to the city. Representatives of city, state, and hinterland debated the relationship between the growing city and a healthy land and disagreed over whether the interests of the grand duchy and its ducal seat were one and the same. Although the unbordered city still seemed un-German and unnatural to those newly caught within the capital's growing circumference, it was precisely by the formal appeal to an urban model that brooked no compromises with the local peculiarities of landscape or power structures that the city lay claim to the special and universal privileges of urban space. Yet this same claim might be seen as destroying the logic of existence of cities such as Oldenburg. The city could no longer be defined by what it shut out—the poor, the foreigner, the marsh, the rising waters of seasonal flooding—because the very notion of the modern city precluded the idea of a stable outside. The remaking of Oldenburg's urban border from the late eighteenth century onward gave visible expression to the conflict between the natural landscape and notions of what constituted a natural or rational city form, ideals that often had little to do with Oldenburg's own place in the economy of marsh, moor, and geest, nestled in the knee of the Hunte River, a tributary of the Weser. Long after it ceased to be the city border, the fortifications land retained significance as a representation of the urban past and of the link between the city and the landscape. In particular, as a relic of the historical relationship between settlement and water, the border represented the interface between landscape and infrastructure, between an elusive notion of unspoiled nature and the technologies of progress and the rationalization of space.

Understanding how the narrative of Oldenburg as a growing city emerged on the ground, in a particular place and landscape, reveals the conflicts of interest and understanding hidden beneath the apparently self-evident emergence of a growing modern city under the influences of modernization, industrialization, and national unification in the nineteenth century. The city's walls, moats, and borders played an important part in the construction of that narrative. In both Oldenburg and Leipzig the old fortifications land remained an important reference point for the urban past throughout the nineteenth century and into the twentieth, even as the physical city walls themselves played a relatively minor role in the cities' nineteenth-century urban developments. As the preservation of visible city borders became more unusual over the nineteenth century, city walls came increasingly to represent the elusive stability of premodern urban boundaries—which could also seem like stagnation—and remained a reference point for later expansions. The city wall as a physical artifact itself could also become a point of contention in the latter half of the nineteenth century.

3 Taxing the Urban Border

The Persistence of Prussian City Walls

In 1818 the Westphalian president, Ludwig von Vincke, wrote to the Prussian chancellor, Karl August von Hardenberg, that in Prussia's newly acquired western provinces, the divisions between city and country had all but vanished. The separation between them, he wrote, had long been exaggerated, since city dwellers continued to make their living by agriculture, and manufactories were as often as not rural endeavors. In the context of this already fluid relationship, recent reforms had definitively loosened the cords of guild and excise tax binding craftsmen to the city. Vincke believed that some cities would survive these changes, if not recover their medieval brilliance—but only a few of the largest and most important residencies and market cities, which, in Westphalia, included only Münster. For lesser cities, distinction from the countryside would soon become an anachronism. All that remained was the fall or removal of the city walls for the last distinction between city and village to disappear entirely. Although his assessment of the changing urban and rural landscape was not inaccurate, twelve years after Vincke anticipated the disappearance of city walls, the Prussian king instead ordered a stop to their "arbitrary" removal, reasserting the importance of the city wall for defense, policing, and closure. Why did city walls persist so long after their purposes seemed to have disappeared?[1]

It was not just in Prussia that the era after the end of the Napoleonic Wars saw the physical restoration of walls. In Bavaria, where the demolition of city walls had been allowed since 1804, the right was again revoked in 1826. As communities and rulers across German central Europe sought security in re-established practices from the past, the city wall provided one very literal way of connecting to tradition. In Yair Mintzker's account of German defortification, he characterizes the period of Restoration after 1815 as one of uneven "piecemeal" changes: walls were variously demolished, reconstructed, and left to decay depending on the whims of rulers and local circumstances. Mintzker identifies broad patterns in these piecemeal changes, based on a three-part typology of cities. In hometowns, walls remained standing when they were perceived to serve the interests of property and community. In fortress cities, walls met defensive needs that were increasingly driven by the state's interests rather than the city's; urban fortifications persisted longer near threatened borders. In the metropolis, walls addressed rulers' anxieties about the need for protections against the citizens themselves—the potentially unruly citizens of growing and politically suspect cities.[2]

When nineteenth-century Germans talked about their cities, needs of taxing, community protection, and defense rarely remained so tidily separated, and nor did they conceive of cities in terms of a clear typology. Instead, the clearest division in the perception of the city wall in this period was proximity to its physical presence. Petitions for the removal of walls and the opening of gates provide a glimpse into the busy lives of nineteenth-century cities and how ordinary city dwellers interacted with the wall as a part of the urban landscape. Bakers, butchers, weavers, and carpenters all saw the wall as it related to their trades: sometimes as a protection but, often, as separating their home from their workshop or as stealing light from the yards and rooms where they worked. The owners and patrons of cafés, inns, and public baths saw the gates, locked at night, as an obstacle to their business or their revelry. These views of the city wall as a boundary were both practical and hyperlocal. Tax officials had an obvious interest in viewing these petitions with suspicion—each window, gate, or door in the wall was another site of potential smuggling. Cash-strapped city governments were mostly worried about having to pay for expensive renovations or removals. The use of fortifications and walls for purposes other than the defense of the city—from recreation to the warehousing of goods to paupers' cemeteries—accelerated their decay. The higher up on the administrative chain and the further from the city itself, the more likely officials were both to object to the removal of the wall on principle and to see the wall as a clear, conceptual boundary. City walls were tools of governmentality, because they made the landscape legible:

asserting conceptual order on the chaotic local environments, especially in newly acquired territories.[3]

Debates over city walls cut across distinctions of place and size, as, for example, in two very different cities: the city of Paderborn, a small Catholic city in Westphalia, and Berlin, the large mostly Protestant capital of Prussia. Paderborn experienced the nineteenth century as a period of decline—of fall in importance, of missing growth, and of a movement from centrality to marginality. Berlin, on the other hand, experienced the same period as one of explosive growth and a shift into the center of German national and urban life. But there are surprising similarities in the ways in which the two cities' walls were invested with meaning, and in the tensions between the administrative interpretation of urban boundaries and the physical realities of city walls themselves entering the second half of the nineteenth century. In both Paderborn and Berlin, the persistence of the city wall gave rise to concerns about how the city was out of sync with its time. As in Leipzig and Oldenburg, city walls became convenient ways to frame interurban competitiveness—though what cities one chose as suitable "peers" for competition could be very telling.[4]

Either looking to the case of a single city in isolation or generalizing the experiences of cities by type or region in historical accounts naturalizes geographies of centrality and marginality, which can make the fall of the wall appear to be the effect of a general dynamic of universal urbanization and the centralization of the nation-state. Differences between cities in region and type were not givens but, instead, were actively constructed by urbanites who struggled to understand their city's place in the world through its participation in the common experience of "tearing down the wall." After 1815 city walls already signified something outmoded. Debates over when and how to remove them engaged ideas concerning history and urban growth and development and expressed anxieties about status and place. When Germans debated city walls they talked about not just community, defense, and commerce but also their understanding of history and synchronicity.

The disjuncture between the importance of the city wall as an abstracted boundary and the material reality of crumbling walls and gates that no one wished to maintain provided the conceptual space for a struggle between local city administrations and the centralized Prussian state over control of the city border, often phrased as a conflict between what kinds of borders were "natural" or "artificial" to the city. This struggle was not usually played out as an easily romanticized battle over the opening of the city or the defense of the urban fortress but, rather, in skirmishes over who had permission to hold the city keys or who would pay to rebuild a collapsed bit of wall. The relationship between the city wall's symbolic importance and its physical preservation and decay shows how the city wall continued to figure as a powerful representational boundary in the discussion of urban expansion and development through the second half of the nineteenth century.

TAXING THE URBAN BORDER

Since the seventeenth century the logic of the Prussian tax system had relied on sharp divisions between city and country, and between indirect and direct taxes. The burden of indirect taxes on consumption, production, and movement fell more heavily on city dwellers, and the rural system relied more on direct taxes, like those on property. This system was not called into question until the Prussian reform movement, led first by Heinrich Friedrich Karl vom Stein and then by Karl August von Hardenberg.[5] Stein first stated the goal of equalizing the legal status of city and countryside in the Nassau memorandum (*Nassauer Denkschrift*), published in 1807. He abolished the old practice of having special municipal tax councilors (*Steuerräte*), whom he believed interfered with the self-government of merchants in the cities, in the former bishopric of Münster in 1803 and throughout Prussia in 1808.[6] Hardenberg initially sought to pursue the same goal with his reform of the Prussian tax system but instead he preserved a clear division between urban and rural taxes. The 1820 tax reform included a milling and slaughter tax (*Mahl- und Schlachtsteuer*) in 132 Prussian cities as an alternative to the rural and small-town "class tax," a yearly household tax of a fee that was determined by the taxpayer's class.[7] These cities included Posen/Poznań, Königsberg/Kaliningrad, Danzig/Gdansk, Breslau/Wrocław, Berlin, Magdeburg, Erfurt, Bielefeld, and Cologne.

The Hardenberg reforms reflected a compromise struck between the desire to rid the tax code of feudal distinctions and the need to raise tax revenues without causing undue ire among the Prussian nobility. Although the milling and slaughter tax reinforced the division between city and country, and so seemed counter to Stein's stated goal, it was part of an ambitious slate of revisions to the tax code that abolished a number of other regionally differentiated taxes. Ridding the tax code of other excise taxes and duties that distinguished between regions and localities, the new milling and slaughter tax had the effect of heightening the visibility of the division between city and country within the tax code as compared to other borders and regional differences. The introduction of a tax on grains and meat throughout the whole kingdom proved impractical in small towns and in the countryside, where individual households took care of their own baking and slaughtering. Instead, these kinds of taxes separated urban space from rural on the basis of a key difference between city and country economies, by taking advantage of the divisions of labor that came with a more urban style of living. However, this distinction was not easily drawn. Some early resentment of the milling and slaughter tax came from small cities, like the Westphalian town of Hamm, where slaughtering and baking were taking place in a more characteristically rural manner, in individual households rather than performed by professional bakers and butchers.[8]

The milling and slaughter tax decisively shaped the experience of the city border in many of Prussia's larger cities. Anyone entering the city had to declare products made with flour or meat—bread, noodles, baked goods, tallow, sausage—and either pay the tax or document its origin from a place where the milling and slaughter tax was not owed.[9] The milling tax affected all grain and legumes brought into the city to be milled, including rye, barley, wheat, and corn, but excluding malt for brewing. The meat tax included beef, mutton, goat meat, and pork. The tax was collected two ways in each case: on raw grains and on processed flour or baked goods, and on meat and meat products or on the slaughter of individual animals. City gates served as checkpoints, making the city wall a suspect landscape of possible smuggling and tax fraud. Bakers and butchers within one-half mile of the city paid both the household-based class tax and the milling and slaughter tax, creating a suburban double burden that became a major point of complaint among residents of suburbs and urban border zones. In some large cities substantial numbers of livestock routinely entered the city.[10] By the 1860s, Berlin had hundreds of slaughterhouses within its borders. The assessment of duties at the already burdened entry points to growing cities introduced substantial practical difficulties, especially since livestock tended to enter cities in large herds, creating particularly malodorous traffic jams.[11]

The number of cities in which the tax was exacted declined almost as soon as it was instated. City magistracies had the option of electing to switch from the milling and slaughter tax to the class tax, but in order to do so they had to demonstrate the switch would not decrease tax revenue. The milling and slaughter tax tended to be more profitable where populations were dense and relatively wealthy. Unsurprisingly, the tax increased the number of bakers and butchers just outside the affected zones and decreased the number within them.[12] The tax on meat and grain came to symbolize the unwelcome burdens that attended Prussian administration. A call for its abolition was among the pamphlets distributed in Cologne when, in July 1830, news of revolution in Paris inspired residents first of Aachen and then of Cologne to protest against the Prussian government. In the same year in the small Westphalian town of Hamm, placards demanded that the tax be lifted and issued threats to the tax collectors. The town of Warendorf experienced similar violent scenes in protest against the tax, demanding it be replaced by the class tax within the city's walls. In 1835 the citizens of Erfurt protested the tax. The former imperial city of Wetzlar saw tax protests throughout the 1830s. By 1838 the number of affected cities was already down to 118.[13]

Protests against the tax continued to rise with prerevolutionary (*Vormärz*) tempers. After years of complaints and attacks on tax collectors, the citizens of Hamm submitted a petition to the Westphalian provincial parliament in 1845 demanding that the milling and slaughter tax be lifted. Paderborn and Bielefeld began investigating the possibility of a switch.[14] Taxes were among those issues frequently taken up in liberal journals like Otto Lüning's *Das westfälische Dampfboot* (The Westphalian steamboat). In its pages, socialist Joseph Weydemeyer identified the milling and slaughter tax as the most objectionable part of the Prussian tax code, asking "can there be anything more laughable than this tiny customs chain [*Douanenkette*] in a land that is in nearly every direction bisected by railroads, obstacles to traffic in a time in which everything strives to accelerate traffic?"[15]

The comparison of the taxes exacted at city borders to the duties collected at state borders struck a particular chord in the years following the creation of the Zolleverein (the German Customs Union) in 1834. This economic coalition, created to ease the burdens of tolls and tariffs, rested on Prussian initiatives since 1818 and protected its economic interests, but it also embodied the push for freer trade and rationalized economic policies.[16] As Customs Union advocate John Prince-Smith wrote in 1845, import tariffs inhibiting commerce had gained new importance in the era of the railroad. "Now," he wrote, "since commerce, so wonderfully liberated from natural shackles is orchestrating a general reorganization of industry, interference with the natural course of development must have the most wide-spread and far-reaching consequences."[17] The discussion of urban borders echoed these claims. Another critic of the urban border taxes wrote: "As we strive to realize the ideal of a free intercourse among peoples, and to awaken through the Customs Union the common interests of the greater society, there remains in the interior itself, at the gates of the great cities, the restriction of an internal tariff; even from the windows of the iron machines we are reminded of a time when the acerbity of the gate keeper was commonly known."[18] Taxes at city gates seemed at odds with the political and economic impulse behind the foundation of the Customs Union. Critics like Weydemeyer found it paradoxical to dissolve state borders through the establishment of new institutions and the spread of railroads, at the same time retaining "in the interior itself" tariffs and gatekeepers at the city edge. The urban taxes were doubly anachronistic: recalling an older form of the city (closed, guarded) as well as an outdated economic system and trade protectionism.

Under increasing pressure, urban border taxes were repealed in many of the remaining cities. Bielefeld and Arnsberg were freed from the milling and slaughter tax in 1846. In 1847 and 1848, it was replaced with the class tax in all of Westphalia and most Saxon cities. Paderborn and Minden had demanded exemptions in April and May 1848. An order of April 4, 1848, laid forth two options: a city could apply for the tax to be lifted and pay their tax another way, possibly through an addition to the income tax; or the city could keep the milling and slaughter tax. On July 7, 1848, a new order ruled that the class tax be introduced in Minden, Paderborn, and Münster, the last three Westphalian cities to pay the milling and slaughter tax.[19]

Although never revived in Westphalia, after a brief three-year hiatus, the new 1851 Prussian tax code preserved the milling and slaughter tax elsewhere in the kingdom. The failure to secure approval for any substantial reforms, especially in the shape of an income tax, made it difficult for the state to forgo the income won from less efficient indirect taxes, like those collected at the city border. The tax continued to irritate Prussian urbanites in the 1850s. Berlin was not exempted from the taxes until 1873, after the city's wall was removed in the 1860s. Contemporaries remarked on the increasing absurdity of its continued collection precisely in many of those cities that seemed to defy control at the border because of their size and growth. The protests gained in volume again in the early 1860s. By this time the tax was even more inefficient than before because of the substantial administrative costs (roughly 13–15 percent of revenue).[20] To opponents, it seemed to be a tax on city growth itself and drew a direct link between the size of a city and the injustice of the tax. "The more a city gains in circumference," one pamphleteer wrote, "the greater its daily traffic becomes, the more substantially it suffers through the milling and slaughter tax." Another common complaint was that it was unfair to the lower classes. Taxing the entry of basic foodstuffs into the city penalized the urban poor, who were forced to rely more on nutritionally poorer foodstuffs like potatoes, because of the increased cost of corn and meat. Largely because of surpluses created by the boom of the early 1870s, the Prussian government decided that the milling tax should end in 1873, and the slaughter tax the year after that, finally ending a structural distinction between urban and rural space that had lasted centuries.[21]

The fight over the milling and slaughter tax was significant for the evolving relationship between the city and the state and for the experience of the urban border for two reasons. First, the debate over the appropriateness of indirect taxes, of which this was the most visible and resented, focused on the suitability of limiting the free movement of goods and people by constructing "arbitrary" spatial borders. Second, imposing a tax on cities (more or less) on the basis of their size emphasized a common identity among cities that otherwise shared very little, and the appearances and operations of which did not obviously follow a common pattern. Taxes exacted at urban borders drew attention to the changing social and economic identity of cities and invited relatively small towns to fight for their rights by participating in a common urban identity with larger cities.

THE "ARBITRARY DESTRUCTION" OF PRUSSIAN CITY WALLS

Collecting urban excise taxes depended on the maintenance of some kind of secure and reliable border between cities and their hinterlands. Borders that had at one time signified urban independence and power became instead markers of state control in local spaces. Crumbling, decaying, and unrepaired walls were not just an embarrassment to cities but also raised concerns about tax fraud. An examination of the difficulties experienced in collecting taxes in diverse cities provides a rich overview of the appearance and experience of the urban border during the tumultuous years of the Vormärz. In some cases, both provincial tax administrators and the government in Berlin perceived the difficulties they experienced maintaining decaying walls, fortifications, and moats to be evidence of local intransigence.

State control of cities' walls and borders led to some incongruous and surprisingly intimate interventions of state power on the level of the urban street. This is evident in the petitions received by the Finance Ministry and the provincial governments requesting permission to open new gates into the city. With the exception of significant urban fortresses and some large cities, most city walls were not unbroken, free-standing circuits constructed of a single piece and design but, rather, were composed of disjointed segments built at different times and patched together or extended over the years. Wall structures were frequently integrated into the urban fabric and building stock, rather than standing isolated from the surrounding environment.[22] They could provide structural support or even a fourth wall for adjacent buildings, whether publicly or privately owned, and could simultaneously serve as a public administrative border and private property boundary. Towers and other habitable portions of the fortifications could be rented out to individuals for use as residences or workshops.

In spite of this complex variety at the ground level, the categorization of walls as a border for state tax collection meant that changes to their structures required approval from outside the city. In a common kind of case, a weaver in the Upper Silesian city of Neustadt wrote to the Finance Ministry in Berlin in 1833 for permission to construct a door between his house and his garden. His house shared a fourth wall with the city and his garden was on the other side, meaning that the urban tax border separated them. His petition was rejected, because of concerns about tax fraud.[23] Besides dividing houses from their backyards, tax walls could also disrupt city business. Weydemeyer and other reformers depicted the disruptions arising from the milling and slaughter tax on an epic scale (steam engines clashing with gatekeepers, for example), but actual city walls disrupted city life in numerous small quotidian ways, separating a small craftsman from his materials, a carpenter from the light he needed to ply his trade, or an innkeeper just beyond the city gates from his customers.[24] In many cases it had been customary for the owners of workshops just beyond the walls to have access to private gates or a special key. When control over the wall was transferred from city authorities to the state, these privileges could be revoked. This is what happened in a contentious case in 1823 in the mid-sized Hessian city of Wetzlar, located in an exclave of the Rhine Province on the Lahn River. Tax officials accused the owner of a mill on the river just outside the city walls of using a gate that had been privately owned by his mill since 1720 for the purposes of smuggling, so they confiscated his key to the city. The miller, a man named Peter Rau who was characterized as "raw" and "recalcitrant" by officials, did not accept the decision quietly. He complained that retrieving the key for the gate from the city tax office took valuable time in emergencies like floods and fires. In his frustration, he attacked the tax officials who kept watch at the gates, ripping the key from one and flinging it at his head. Even after this event, he sent a further petition all the way to Berlin. This was unsuccessful. A long admonitory letter from the Finance Ministry rejected his petition and lectured Rau that he must learn to "keep within the limits of lawful order."[25]

In cities subject to the milling and slaughter tax, the freedom to pass in and out of the city was a privilege not granted to just anyone. Both petitioners and officials continued to associate the necessity of controlling movement in and out of cities with concerns about the trustworthiness of city dwellers, but the years of French rule had raised expectations in occupied cities for greater freedom of movement and had begun a process of opening gates and eliminating border controls that was difficult to reverse. Among petitions submitted to the Finance Ministry involving urban walls, the most common type came from an innkeeper seeking permission to construct a new gate in order to ease the passage of potential customers between the city and his inn, placed, as inns and guesthouses often were, just beyond the wall. Also common were petitions from millers, tanners, and other small craftsmen who wished to ease the transportation of wood and other materials to and from their workshops.[26] Individuals appealed to privileges granted earlier, sometimes under French administration, that were now seen to undermine the financial interests of the state. Petitions submitted to Berlin often appealed to ideas of class, status, and respectability to justify their claims.[27]

Prussian administrators often saw the unnecessary removal of city walls in terms of local obstinacy and disobedience. In 1818 and 1821 the provincial governments of Silesia and Pommerania banned the removal of city walls.[28] In 1829 Interior Minister Friedrich Freiherr von Schuckmann, War Minister Karl Georg Albrecht Ernst von Hake, and Finance Minister Friedrich von Motz advised King Friedrich Wilhelm III to support a ban of the "arbitrary destruction of cities' walls." Making an exception for those cities with walls that were already too dilapidated to maintain at a reasonable cost (a standard that would itself become a source of conflict), they were concerned that cities had been removing their walls and fortifications for reasons that were selfish, trivial, or not in the interest of the Prussian state. Some cities, they asserted, wished merely to use the stone and other material won by dismantling fortifications for other construction projects. Others acted "only to achieve the expansion of their cities and their connection to the suburbs." This was apparently insufficient justification for taking down a wall.[29] Worst of all, some cities removed their walls out of a "mere love for change and improvement."[30] The three ministers advised that cities were ignoring orders from the Prussian government to cease removal of their walls and were supporting their right to do so by arguing that the walls belonged to the city and not the state. In their view, the increasing rate at which cities were undertaking these unacceptable projects indicated that a new law was needed, even though this "arbitrary destruction" should have already been understood as prohibited by the Allgemeine Landrecht (ALR), the Prussian universal common law code, which stated that "so far as an object's preservation has a significant influence on the preservation and advancement of the public good, to this degree the state is entitled to prohibit its destruction or demolition."[31]

In accordance with this advice, on June 20, 1830, Friedrich Wilhelm III passed a new law to make it more difficult for cities to remove their walls. The law stated that the "arbitrary removal of city walls, gates, towers, ramparts and other structures intended to close or defend the city could not be allowed for constabulary, military, and financial considerations." Even in cases of extreme dilapidation, any demolition of a city's border defenses required state approval.[32] This seems to have been the case regardless of who owned the wall, even if it were city property. The law failed to clear up the question as to who was responsible for paying for the maintenance of city walls, stipulating that financing always be decided on a case by case basis, although it did establish a new system of standards by which to determine whether a wall was worth maintaining. After 1830 petitions for opening cities' borders were considered by the threefold criteria of whether the closure of the city served constabulary, military, or financial purposes.[33]

Were cities really dismantling their walls against the wishes of the state? There is much more evidence of passive refusal by cities to maintain or repair their city walls. Taking the example of Westphalia, city officials in Paderborn and Münster clashed with tax administrators over financial responsibility for the walls, but underlying these disagreements was also the question of local control. Münster is a particularly interesting case because of the intermingling of military, social, environmental, and economic concerns. In the 1830s there was an extended debate over the closure (*Stadt-Verschluß*) of Münster that gives a good sense of what was at stake after the passage of the 1830 law. The city petitioned for permission to remove one of its bastions. Ownership of its fortifications was transferred to military authorities after the milling and slaughter tax was eliminated in 1848. In Westphalia, only Hamm was accused of actively undermining the seal of its wall, but it is unclear whether this claim was justified. In 1841 tax administrators reported that Hamm had long refused to pay to maintain its own "gate closure" (*Thor-Verschluß*) by neglecting necessary—and expensive—repairs. The city magistracy asserted that the city itself had no interest at all in the closure of the city, so should not be responsible for payment.[34]

In the quarter century following the introduction of the milling and slaughter tax (1820–1845), at least six Westphalian cities—Warendorf, Bielefeld, Paderborn, Münster, Hamm, and Minden—reported difficulties collecting taxes because of the physical state of their walls.[35] This represents fully half of the twelve Westphalian cities that were then required to pay excise taxes instead of the class tax.[36] They ranged from the small (Warendorf) to the large (Münster), and from the heavily fortified (Minden) to the weakly defended (Paderborn). In May 1820 district government officials in Münster wrote to the Finance Ministry in Berlin reporting that Warendorf was plagued by border smuggling, especially during the nighttime, because of the insufficient closure provided by its walls. In July, they registered similar complaints about Münster itself, requesting and receiving funds to strengthen the walls and gates.[37] In September 1822 the district government in Minden complained that in Paderborn the poor condition of the walls made it easy to sneak in flour.[38] Although Bielefeld was mostly enclosed by walls and, even where these had deteriorated, by wet moats, both the walls and moats were in such poor shape that "it was possible nearly everywhere to cross [the border] with little effort" to smuggle in goods without paying the appropriate taxes.[39] Maintaining old urban fortifications for the purposes of tax collection was expensive and labor intensive.

In Hamm, where there were early and vocal objections to the milling and slaughter tax, the problem was not only the natural decay of the walls but also miscommunication between different levels of the bureaucracy. In September 1830, shortly after the summer unrest there, the provincial tax director in Münster reported to the Finance Ministry that the district government in Arnsberg had approved the removal of one of the city's four guarded gates, locked every night, because of its advanced state of dilapidation and had "therewith abrogated the closure of the city" without consulting with the tax administration first.[40] In each of these cases determining who owned the wall and who was responsible for its maintenance was a source of ongoing confusion and conflict. Military and financial considerations were the sole purview of the state, but the interests of the police could concern the municipality as well, so maintenance of a wall useful for reasons of "policing" could be the financial responsibility of the city.

Whether or not cities' desire to open their borders and achieve better connection to their suburbs led to frequent disobedience and the removal of walls and gates necessary for tax collection, neither Friedrich Wilhelm nor his ministers considered urban expansion and "improvement" to benefit the state. For one, the reasons for city growth were not themselves seen in a positive light. The pamphlet literature on pauperism in the 1830s blamed population growth for exhausting resources and sapping national strength. Conservative Vormärz governments urged emigration abroad and strengthened marriage restrictions on the lower classes in an effort to stem population growth, because it was seen as a threat to the state. Moreover, precisely those reforms that effectively eliminated the legal and economic borders of the city were routinely blamed for the increase in pauperism. In contemporary investigations, the growing numbers of impoverished and out-of-work artisans were, for example, blamed on freedom of trade, a reform that was seen as undoing the justification for the closure of urban space.[41] Capital cities that functioned as administrative centers and royal residencies were the exception. The problem of pauperism was relatively more severe in industry and trade cities than in residencies and administrative centers. In capital cities, growth could be an expression of, rather than threat to, control from the center. City walls and fortifications were also involved in public discussion of pauperism because they were prominent public works. Both their extension and their dismantlement could be used as work creation schemes.[42]

Although the ostensible importance of the urban border to the Prussian state was as a point of tax collection, the 1830 law banning the "arbitrary" destruction of city walls demonstrated a reluctance to give up the closure of urban spaces for reasons that went beyond the convenient collection of taxes. The city wall was a locus of anxiety about the status of the city and the relationship between the city and the state. A law that treated all city and town walls in the same way was incapable of addressing the tremendous complication of the situation on the ground, but it was a necessary corollary to tax laws that treated urban space as consistently different from rural space, as separate zones for economic and social organization. From the Prussian capital of Berlin to the small Westphalian city of Paderborn, concerns about the delayed destruction of city walls played a key role in understanding and constructing the relationship between centrality and peripherality within both urban and Prussian geographies.

PADERBORN'S NEGLECTED WALLS

On the eve of its removal in 1881, the wall that still enclosed the small Westphalian city of Paderborn was not a proud piece of the urban landscape. In the opinion of most everyone, in the city and out, it should have been removed decades before. This wall did not recall past military victories, nor did it, in its current diminished state, easily bring to mind a now distant heritage of urban independence. Instead, Paderborn's crumbling wall highlighted the city's failure to keep up with the process of modernization and growth that characterized the nineteenth-century city. It reminded Paderborners of their city's failure to live up to earlier promise. By 1880 the practical difficulties involved in keeping the wall in good repair and in managing, cleaning, and policing the incompletely drained moats adjacent to it had become a symbol of Paderborn's difficulties as a city and its marginal position within the German urban landscape. When the wall finally did come down, Paderborn's mayor represented the event as a hard-won victory of the city over the state—but less for the sake of growth than to shed an unnecessary financial burden.[43]

This was the anticlimactic end of a decades-long battle between the city and the Prussian state over the status and significance of the city within the state, and of its walls as a point of financial and political control.

Almost eighty years earlier, when the prince bishopric of Paderborn was dissolved and incorporated into the Prussian state, Paderborn had a long history as both a religious seat and a city with an independent spirit, whose citizens came into frequent conflict with its ruling clerics.[44] The town sits in a deep ravine of the Teutoburg Forest. Its landscape is dominated by three features: the tangle of streams dotted with footbridges and waterwheels where the warm waters of the Pader River originate; St. Liborius Cathedral, which sits perched above the springs in the city center; and the city's twelfth-century walls. The medieval walls, built during a period of economic growth and civic health, remained unchanged for centuries, representing the rough limits of the city's settlement until late in the nineteenth century. The high points of the city's urban independence came in 1254, when it joined the Hanseatic League and enjoyed a brief period of economic flourishing as a city known for beer production, and in the sixteenth century, when the majority of its citizens briefly converted to Protestantism in protest against clerical abuses of power. Defeated in 1604, the city lost its right of self-governance to the Prince Bishop.[45]

Although once an important religious and administrative center, by the early nineteenth century the city of Paderborn had fallen on hard times.[46] The city lost its status as a fortification in 1787, after which its ramparts and walls, already deteriorated, fell into a state of advanced decay.[47] Descriptions of the city's dilapidation were further exaggerated by anti-Catholic sentiment. A traveling Prussian official who had been favorably impressed with Oldenburg's enlightened environment observed in 1802 that Paderborn had a "very miserable appearance." He described the city as decayed and backward, burdened with a superfluity of churches and cloisters. Deploying common anti-Catholic tropes, he wrote that the promenades outside the gates filled with the "buzzing of a terrible swarm of French priests."[48] Although reduced in population, wealth, and importance, the city continued to have a lively associational life through the beginning of the nineteenth century. Nonetheless, secularization had dealt a hard blow to Paderborn's importance and self-sufficiency as a city. When it was incorporated into the Prussian state, it lost the right of communal self-government: municipal officials were now appointed instead of elected. The city did not regain the right to self-government until the passage of a new municipal order in 1836.[49]

Negotiations over Paderborn's city wall from the institution of the milling and slaughter tax in 1820 until the wall's final removal in the 1880s centered on two persistent themes: first, the city's relationship to the state in terms of local control and civic privilege; and, second, the disjuncture between policies that treated the wall as a clear and more or less impermeable border, in theory, with the reality of the partial and crumbling wall itself. The presence of the city's old walls in this dilapidated state through most of the nineteenth century served to reinforce prevalent stereotypes about Paderborn as a small backward town as it came into contact with the Prussian state bureaucracy (see figure 3.1). The antagonism between Paderborn and Prussian officials over the city wall peaked around 1830 and again in the early 1850s, times when the relationship between city and state were already tense. Prussian officials both in Berlin and in the district and provincial capitals of Minden and Münster became frustrated with the city's recalcitrance when the wall needed improvements in order to collect the milling and slaughter tax. Later, Paderborn's administrators again thwarted Prussian officials' plans by their unwillingness to cooperate with plans to remove the city wall in the 1850s. Even though the wall remained an attenuated site of potential social and political control, to be removed in the name of urban "improvement," Paderborners themselves showed relatively little interest in its removal.

The first disagreement between the city of Paderborn and Prussian administrators concerned the wall's status as a tax border in the late 1820s and early 1830s. This was a time of severe economic stress in the predominantly agricultural land of Paderborn. A string of bad harvests brought heightened tension over the collection of taxes.[50] The introduction of the milling and slaughter tax in 1820 made a tighter closure of the city necessary for effective enforcement at Paderborn's gates. This required substantial repairs and modifications to the existing structure. The initial battle over Paderborn's city wall was waged not over whether the enclosure would remain but, rather, over who would pay for its upkeep: the cash-strapped city or the cash-strapped state. The issue was a financial one, but Prussian tax authorities thought of and justified their requests in terms of Paderborn's history of lost status and independence.

In 1822 the regional government in Minden reported that the decayed condition of Paderborn's walls and its many unguarded openings enabled easy smuggling of untaxed flour. They promised the Finance Ministry in Berlin that Paderborn's numerous unpoliced entrances were in the process of being closed and that the revenue from Paderborn's mills along the river outside the city would soon increase, but city officials claimed no responsibility for the maintenance of its walls. City and tax officials entered a stalemate: tax

3.1 Paderborn's city wall just peeks out from behind the trees surrounding the city in this view from 1842. Heinrich Wilhelm Teichgräber, *Paderborn*, 1842, Lithograph, 18 x 12 cm, Heinrich Heine Universität Düsseldorf, http://digital.ub.uni-duesseldorf.de/urn/urn:nbn:de:hbz:061:1-137154/.

authorities outside the city repeatedly complained about the inadequate "closure of the city" and Paderborn repeatedly refused to do anything about it.[51]

Further investigations in 1827 and 1828 revealed that the difficulty involved in collecting taxes in Paderborn stemmed from two sources. First, natural processes of decay left the walls crumbled and easily breached. The struggle between city and Prussian officials was a race against the physical decay of the wall itself, which continued to crumble while correspondence flew between Paderborn, Minden, Münster, and Berlin. Second, special permission had been given in earlier years for the construction of a number of private entrances for mill owners located outside the city, which had since become locations for tax fraud. The private entrances were shut, in one case over the strenuous objections of the owner who complained that without his gate he was "outside the ring wall entirely enclosed on an island."[52] This process of enforcement reveals how the logic of the tax intersected with the particular hinterland of branching streams outside the city.

Although they cooperated with these requests in 1828, the city director and magistracy of Paderborn remained adamant that, although the city owned its walls it was not responsible for their maintenance or restoration.

The Finance Ministry asserted that the upkeep of the city walls had constabulary (*landespolizeilich*) as well as financial urgency.[53] The squabbling over the wall reflected deeper tensions over Paderborn's status within the Prussian state. The question concerned which privileges remained of the city's past as capital and residence of an independent land. Paderborn's claim to exemption was one based on precedent. The city government argued that, since the city had never before paid for the maintenance of its walls, even when it had been the residency of the prince bishopric of Paderborn, it should not have to pay for them now. The general director of taxes in Berlin argued that prior precedent had no relevance since Paderborn "had ceased to be the fortress, residency and control point of an autonomous land." The Finance Ministry insisted that any claim of exemption from the obligation must be based on a legal title specific to the city of Paderborn and not "on the general principle of civic freedom and laws, according to which the city believes it can do what it wishes with its walls." The matter was eventually decided in court when the provincial tax director brought a complaint against the city. The Royal Prussian High Court ruled on the side of the city in 1833.[54]

Although the city had been reluctant to cooperate with measures to enforce the milling and slaughter tax, this had little to do with an objection to the tax itself but, rather, remained a question of the city's legal rights and the state's financial obligations. From the late 1830s through the early 1840s, the Paderborn city council went back and forth on the issue. The city council in 1837 and 1845 recommended that the milling and slaughter tax be lifted but in July 1846 recommended against its replacement with the class tax. On March 16, 1848, the council members switched positions once more and declared support for the change. On April 4, 1848, the tax was finally lifted. The only city tax to be collected at the gates, a small fee used to pay for city street maintenance (*Pflastergeld*), had already been lifted in 1841.[55] With the elimination of the milling and slaughter tax in 1848, the last financial justification for the preservation of Paderborn's wall disappeared. Nonetheless, it took more than another thirty years for city and state to reach an agreement on its removal.

The following decade brought a number of changes to Paderborn's urban landscape as the city underwent a brief period of rapid growth. In 1850 Paderborn received its first railroad connection to the nearby city of Hamm. Critics of the milling and slaughter tax envisioned a dramatic confrontation of walled city and steam engine, but when the first railroad reached Paderborn, it kept a safe distance. The city's first train station, completed in 1859, was placed outside the city, far beyond the Western Gate. Its construction fueled population growth, attracting both skilled and unskilled labor. In 1855 the city received its first gas street lighting. In 1859 the city built a hall on the filled-in moats outside of Rose Gate for a new wool market, which quickly became a center of the regional textile trade.[56]

The growth and dynamism of the city in the 1850s combined with the elimination of any financial rational for the preservation of a closed border might lead one to expect pressure for the removal of Paderborn's city wall. The early 1850s were also, however, a period of tension between the city of Paderborn and Prussian officials, as Jonathan Sperber's study of counterrevolutionary efforts in Westphalia has demonstrated. In April 1851 officials feared that the public celebration of the feast of the diocese's patron, Saint Liberius, would be transformed into a riot by the railroad laborers who were at work outside of the city, especially given Paderborn's known "democratic tendencies." The district government in Minden dispatched a secret investigator to keep a watchful eye on Jesuit activities, and the creation of a royal police inspectorate in 1855 subjected the municipal police force to the direct control of a resident government commissioner.[57] It was the combination of Paderborn's Catholic culture and the first stirrings of industrial development in the "backward" city that inspired such concerns.

In this atmosphere of suspicion and heightened oversight, Prussian officials called on conflicting narratives when it came to the city wall. On the one hand, they portrayed themselves as bringers of "light" and "air" to the darkly walled and airlessly crowded city with its medieval layout, evincing frustration that the city government was not itself more interested in removing the walls for the sake of health and improvement. This supposed self-interest was used to justify pressuring the city to take over ownership and responsibility for the crumbling and cost-intensive structure. Paderborn was visited by its most serious cholera outbreak in 1851, in which 169 people died. During cholera scares, public health orders demanded the clearing of debris and filth from the city streets. The result was the dumping of large amounts of waste just beyond the gates. Repeated cholera scares intensified the walls' association with the ill and the foul.[58] Furthermore, the wall had become so dilapidated that it was itself seen as an immediate health threat to those who lived nearby. The provincial tax director Karl August Göring framed the urgent need for repair as a health issue that concerned especially the poor who lived near the wall. "Apart from the fact that the city has a great interest in being able to freely dispose of [its walls]," he wrote that their removal was also to be desired because they "poisoned the air" of the adjacent "narrow allies inhabited by the poorest classes." An 1851 investigation by a royal commission composed of both civil and military representatives emphasized urgent health reasons for removing the wall.[59]

On the other hand, Prussian officials in Berlin remained reluctant to give up the city's closure entirely as long as they viewed Paderborn as an uncooperative city. From the central perspective of the capital, the wall persisted as a potentially useful tool for oversight and control as well. Although initially reluctant to admit that the wall had any architectural or historical value, administrators in Berlin further cast themselves as protectors of the prince bishopric's legacy against the unwillingness of the city to care for the five towers that were part of the wall and deemed worth saving for reasons both practical and aesthetic.[60]

The same 1851 investigation made only very conservative claims for the military and constabulary use of the wall, arguing that its most significant potential role would be in oversight by the local police (*Lokalsicherungspolizei*) but not by Prussian state police (*Landespolizei*). While regional administrators discussed questions such as health and improvement, in Berlin the wall took on a different symbolic significance as a site of political control. In 1852 the ministers of war, finance, and the interior suggested that a three-foot breastwork be preserved, instead of removing the wall entirely. Where the wall was ceded to individuals for use as a house or garden wall it would remain at least seven or eight feet tall and without doors or windows.[61] Friedrich Wilhelm III received these recommendations, but he did not share his advisors' confidence that the city wall had become obsolete for the purposes of controlling movement into and out of the city. Taking an even more cautious approach, he granted his approval for a very partial removal of the wall, reducing it to a height of between four and one-half and six feet, only where this reduction was made necessary by the wall's poor condition. No need, went his reasoning, to eliminate the possibility of closing the city entirely, if such were not necessary. He noted that this closure could be useful, for considerations both military and constabulary, a judgment that seems out of touch with the physical condition of the wall itself.[62] Most of the damaged parts were removed or decreased in height to lessen the damage of falling stones. Because the city had refused to pay for the wall's removal, it lost the right to give permission to make even minor alterations to its structure.

While these negotiations proceeded, other more immediate decisions about the wall confronted the city. In 1855 fifty-eight Paderborners petitioned the mayor for the placement of a new gate between Western and Neuhäuser Gates and across from the city prison, where an old opening in the wall had been bricked up long ago, in order to ease traffic flow. The project, they argued, would be virtually free, since the gate was already there. The bricks simply needed to be knocked out. The mayor supported their request, but Paderborn's district administrator (*Landrat*) expressed reservations. The Landrat, appointed by an assembly of notables to govern the district, had wide-ranging powers including policing and taxing. The Department of the Interior in Minden approved the gate's construction in 1856, though it was to be wide enough only for pedestrian traffic.[63] Following approval, the Landrat maintained that the new gate should be "closeable" (*verschließbar*), while the city's mayor argued that this was an unnecessary precaution. The Landrat was concerned that too easy entrance into the city center would invite in those prostitutes who were known to offer their services just beyond the city gates. A closeable gate was in "the interest of morality." The mayor replied indignantly that the city gates no longer served as a restraint on passage. Besides, he pointed out, the crumbling wall itself provided little obstacle to entrance. The city's morality would not be newly threatened by the "entrance into the city of the dissolute strumpets who rove the promenade" since they could already enter by any of the other gates, which were no longer guarded or closed overnight. The city got its way over the Landrat's objection when, in July 1856, the Interior Department in Minden supported the mayor's petition.[64]

The discussion over the new gate in 1856 made clear that the city border still symbolized social and moral boundaries.[65] The mayor never disputed that undesirables lurked outside the city. Rather, he stated that keeping them out was an elusive goal, and one that Paderborn's deteriorating city wall was not up to fulfilling. Clearly, a wall *ought* to be able to keep people out, but this was in many cases no longer true. The classification of something as a city wall and its resulting administration from afar, by the state instead of the city, divorced overall policy from material conditions on the ground and so emphasized the symbolic significance of the city wall as a border. The gap between the actual condition of Paderborn's wall and the official perception of it as a functional border again caused trouble a few months later, when a royal police inspector complained that the new entrance "in its current condition has little similarity to a gate." The entrance "presents the eye of the passerby with a thoroughly unpleasant appearance" and even made passage through it difficult.[66] From the inspector's description, it seems likely that the residents of the adjacent neighborhood had, when given permission, simply knocked through the bricks so that they could clamber through: a practical, if aesthetically unappealing, solution. Locally, the wall had been treated as an obstacle to daily routines and pathways rather than as a formal border.

In the 1830s and 1850s, city officials in Paderborn remained relatively passive in the discussion over what was to be done with the wall. Although some individual citizens petitioned for the opening of new gates and passages into the city, the city government remained reluctant to pay. The city first took an interest in projects to remove the wall or open up new gates after the completion of the city's first train station in 1859. In 1860, Paderborn's mayor took the initiative by asking for permission take down part of the wall so as to allow a new row of building sites along the promenade. In 1864 a new pedestrian gate was added to ease traffic. This was followed by the construction of several new gates.[67] But even in 1871, only 300 people out of a population of 13,896 lived outside Paderborn's walls. Although the city had again refused state offers to take over ownership and control of its wall in 1868, the pace of change in the landscape along the wall quickened in the late 1860s and early 1870s, with the removal of Neuhaus Gate and the breakthrough of additional passageways through the wall.[68]

The mayor began to seriously consider petitioning for permission to take down the city wall in 1877, motivated by its decrepit state and adjacent swampy ground.[69] In 1880 Mayor Franz Georg Franckenberg and the city council recommended that the wall be acquired by the city and removed. On September 2, 1881, Franckenberg signed a contract taking ownership of the wall on behalf of the city. Work began in 1885, carried out by "poor transients with the help of city workers," and continued until 1896.[70] Isolated stretches of the wall remained standing, still part of the city today.

City historians in Paderborn, as elsewhere, have tended to see close ties between the wall's eventual fall, the city's late growth, and the opening of Paderborn's urban frontier after a relatively stagnant nineteenth century. Klaus Hohmann, an architectural historian, writes that "the removal of the gates and walls was the [beginning] of the city's expansion," which, after a slow start, accelerated around the turn of the century, as dense settlements appeared outside the wall ring, along axial roads and especially in the area around the city's train station.[71] There is no real reason why this should be true. The walls themselves, and even the narrow promenade along them, did not take up much space and did little practically to hem in the growth of the city. Instead, the causal arrow points the other way. The impetus to finally remove the wall came from Paderborn's growth in the last decades of the century, when the wall's presence seemed notably out of step with the city's identity. The long afterlife of the Paderborn city wall did, however, reinforce a familiar story about Paderborn's status as a backward place, left behind by the main currents of both economic and political developments in the nineteenth century. This narrative arose from Paderborn's identity as a small Catholic city in a large, centralizing Protestant state, and the long history of antagonism, from secularization to the culture war (*Kulturkampf*) of the 1870s, between the city and Prussian officials. This story line was established early on, but moments of tension over the city wall coincided with moments when concerns about Paderborn's loyalty to Prussia and its suspicious "democratic" character peaked, in the early 1830s and early 1850s.

Even when neither political nor economic centers of major importance, small- and mid-sized cities nonetheless participated in many of the same processes and dialogues about urban growth as did the larger cities. Dwellers in relatively stagnant mid-sized cities developed narratives to explain their lack of growth, which are themselves a significant piece of the story of the city's new identity in the nineteenth century. After 1848 Paderborn's old urban borders held a twofold significance. On the one hand, they symbolized a long-lost independent past and local identity. They were a physical reminder of religious and political significance that had been eroded during years of Prussian rule. On the other hand, Paderborn's border is also the story of a missing frontier. The old wall, so long seen as obsolete, ceased to be the functional and visual border of the city only in the twentieth century.[72] The late nineteenth-century assumption that rapid urban growth was natural, inevitable, and organic was made from the perspective of "successful" central places that flourished in an era of industrialization and population growth. To fully understand the birth of an urban frontier myth, one must also take into account cities where this did not happen, cities that were, so to speak, "left behind." Yet, the picture is more complicated. A small city such as Paderborn was not simply excluded from the processes of urbanization and the collective experience of growth and the falling of their walls. Instead, the process by which those walls came down became a way of understanding the city's relationship to a larger geography of progress differentiated by local conditions.

REMOVING BERLIN'S TAX WALL

It is difficult to imagine a German city more different from Paderborn than Berlin.[73] In the century before its tax wall came down in 1865, Berlin had grown from a relatively obscure backwater and garrison town to be a bustling metropolis, the capital of a flourishing kingdom, and an industrial center. Looking back on those decades of rapid growth, the contemporary observer Robert Springer wrote in an 1860s guidebook to the city that Berlin's wall was falling "victim to the metropolitan spirit," a quote often repeated in city histories such as that by David Clay Large.[74] Springer's quote implies that the very nature of a modern growing city made the presence of a wall untenable.

A deeper look into contemporary accounts and records reveals that Berlin had more in common with Paderborn than one might think. Here too the city tax wall finally came down, neglected, already crumbling, and too late, reinforcing narratives established earlier about Berlin as the backward capital of a latecomer state.

The walls at question here are not the urban fortifications brought down in the eighteenth century but, instead, the much more modest border that replaced these defensive structures for the more mundane purpose of collecting taxes and carrying out gate checks. Berlin's eighteenth-century defortification led to the construction of a new tax wall around the city's expanded municipal area.[75] But only a few decades later, this wall came to signify the setbacks and embarrassments that plagued the city in the first years of the century—the administrative severing of the city from its environs, the passage of a tariff law that required the exaction of excise taxes at the city border, and limits placed on its right of self-government—while the large size of the walled area allowed the city to preserve a partially rural character well into the nineteenth century (see figure 3.2).[76]

The first significant move toward dismantling the wall did not come until the 1840s. It was spurred on by three factors: first, the desire to develop new transportation networks around the city; second, the petitions and complaints of those who lived near but beyond the wall; and, third, traffic holdups (though this would not become a serious issue until the 1850s). In response to a petition from residents for the removal of the wall on behalf of the fervent "desire to build" (*Baulust*) in 1847, the Ministry of the Interior acknowledged that taking down the city wall on the west and southeast sides of the city would be desirable but concluded that the project could not be realized as long as important excise taxes were still to be collected at the city's edge. Nonetheless, over the following decade, the walls were removed or shifted outward in a number of places because of local exigencies.[77]

In February 1848 Friedrich Wilhelm IV called for a further consideration of a possible removal of a portion of Berlin's walls in light of plans to build a tramway around the city. These plans were laid aside after the outbreak of violence in the city, when crowds gathered in the streets in the following weeks demanding reforms that included a constitution and freedom of the press. Hundreds died on Berlin's barricades in March. Following this tumultuous turn of events, the 1850s proved a decade of dissatisfaction and resentment. Residents who lived beyond the wall saw themselves as unfairly burdened by the excise taxes exacted at the city gates.[78] In spite of pressure from residents of Berlin's hinterland, the specter of 1848 revolutions continued to haunt

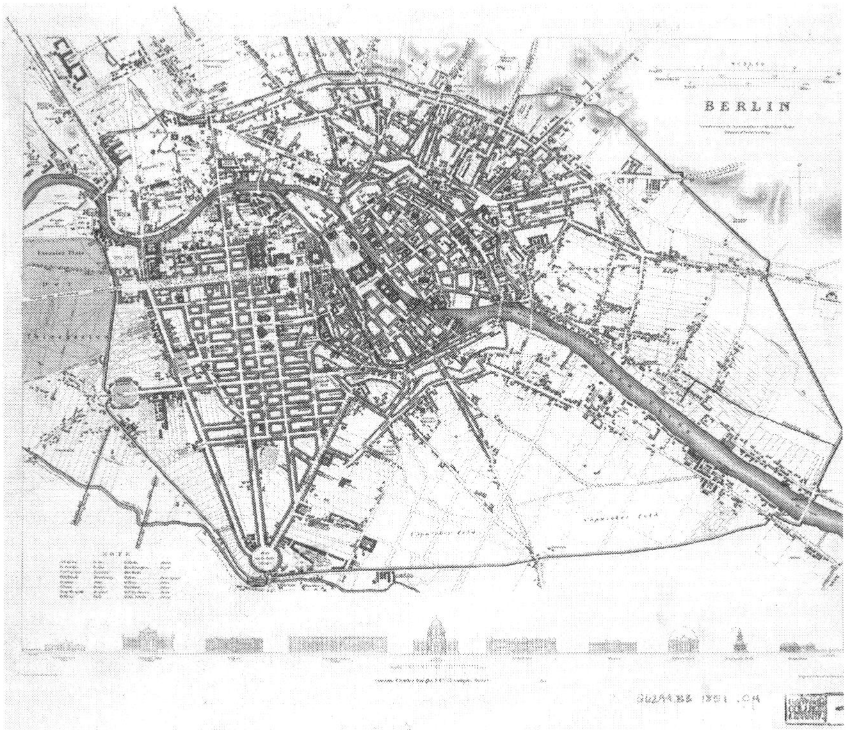

3.2 This map of Berlin in 1851 shows the undeveloped space in the east and south of the city that was still within the tax wall not long before its removal. W. B. Clarke, *Berlin* (London: Charles Knight, 1851), Harvard Library.

proposals to remove the wall. In the 1850s further petitions and plans for the possible removal of the walls were rejected in part for financial reasons but also with reference to the possibility of a second 1848. Officials feared that the absence of any available means to close the city by controlling its entrances would have serious consequences in the case of urban unrest.[79]

Anxiety about revolution on the city's borders reflected the social geography of Berlin. From the 1820s onward, large tenements (*Familienhäuser*) had sprung up outside Hamburg Gate. Here, members of the working class who could not afford to live within the city found cramped quarters beyond the walls where land was cheaper. In the 1840s the contrast presented between the wealth within the city and the poverty beyond its gates seemed an especially stark representation of the way in which pauperism of a new kind and scale threatened to overwhelm the city's social and economic structures in the throes of early industrialization. Urban growth appeared as the result of both excess luxury and abject poverty. The city wall was the border between the two worlds. The question of the city's future development was whether Berlin's character would remain that of the city within the walls or whether, instead, a new Berlin would grow from those who had been expelled from the city and lived beyond the gates.[80]

In the middle of the century, a greater interest in the city's past development, and especially its spatial evolution, accompanied new attempts to control and plan Berlin's future growth. City planners and reformers feared that Berlin, beginning with the initial removal of its fortifications, had never dealt properly with its borders, leaving the city haunted by the walls' partially repressed and ghostly presence: a conflicted and confusing city of both disconcertingly rapid growth and unnaturally perpetuated restrictions on that growth. Because it had grown faster and sooner than most other German cities, Berlin was denied the opportunity to construct a proper ring-shaped promenade. Instead, the marshy land of the former ramparts became among the city's least attractive streets. Without the visible reference point of a promenade or ring road marking the border between the old and the new, the city lost touch with the supposedly natural growth pattern of radial streets and concentric rings.[81]

The city code of 1808 artificially impeded Berlin's development by disrupting its relationship to the hinterland. The code effectively cut Berlin off from its suburbs by giving the city magistracy control only over the land contained within the ring wall that had replaced the city's fortifications in the eighteenth century.[82] This wall enclosed much empty land, but it seemed to many later commentators that limiting urban administration to within those walls thwarted Berlin's healthy growth. This act set up Berlin's fate as a site of conflict between two systems of spatial and social order: that of the state and that of the city, with serious and long-lasting consequences for both Berlin and Germany.[83]

The first half of the nineteenth century was a period of missed opportunities for growth. Some commentators focused on the 1808 city code, and others highlighted the significance of 1848, when the removal of Berlin's ring wall was first considered. The end of Berlin's traumatic separation from its hinterland could be dated to 1858, with a proposed new city plan; to 1860, when extensive areas beyond the ring wall were incorporated as city land; to 1865, when the wall was finally taken down; or to 1871, with the birth of the German Empire and Berlin's subsequent boom in growth. Even after most legal and logistical restrictions on growth had been lifted, nineteenth-century commentators feared that certain remaining habits of mind placed further artificial limits on Berlin's urban development. The trauma of the long-vanished fortifications manifested itself in a kind of "inherited illness," in which the city tended to both artificial density and a too rapid, debauched form of growth. This "illness" hampered attempts to remove the city wall and slowed the city's subsequent growth into the 1860s and 1870s.[84]

Just as in Paderborn, Berlin's city wall bore a multifold significance. It represented a superseded border and a long outgrown and unnatural restriction on the city's commerce and growth. It symbolized a conflict between city and state administration. It also evoked the emergence of new and dangerous divisions between rich and poor, included and excluded. Although the eventual removal of the city wall was welcomed on all sides (except perhaps by the tax collectors), the way in which the significance of this event was understood and described reflected the need to reconcile conflicting models of urban space. Tearing down the city wall, as a celebrated event, seemed both a natural consequence of urban development and a subversive victory of the periphery over the center.

After the 1860 expansion of Berlin's municipal area, attention shifted from the city border to the presence of the physical wall itself. The wall increasingly came to seem an arbitrary barrier between the city core and its suburbs. Negotiations over the wall's possible removal were complicated by confused jurisdiction and tension between the state and city governments. Although it had garnered little attention before 1860, within a few years of the municipal expansion, the wall quickly became a joke, its continued presence a subject of public mockery. Artificially preserved by bureaucratic red tape, it became a seemingly permanent and embarrassing element of the urban environment.[85]

On July 1, 1865, the city's tax border was moved out from the existing city wall to the new municipal boundary, though it would be a number of years before the tax was eliminated entirely. This shift divested the city wall of its most significant remaining purpose. On June 20, 1865, in anticipation of the imminent change, King Wilhelm had authorized the wall's gradual removal—contingent upon the cooperation of adjacent landowners—once the transition had been made to the new tax boundary. The move had been in discussion for years, but in spite of the long preparations and negotiations among city, state, and landowners, progress after the king's decree remained slow. The city sought to manage the wall's removal as a routine clearance project, not as the momentous opening of a border. The most immediate result of the announcement was to bring the wall into the public eye. The planned removal quickly became a common topic discussed in Berlin's local press. But the story told there took a different form from the one anticipated. Officials had underestimated popular excitement over the wall's removal, as well as the impatience that the slow pace of this—admittedly large—public works project would drum up.

Shortly after the June 20 decree, local papers began to report the mysterious appearance of new openings in the city wall. As negotiations with neighboring landowners over the details of funding and construction began, "a troop of little revolutionaries" took more immediate action, setting about the wall with their own hands, and breaking through passageways of their own.[86] Most of these new "gates" were just big enough to crawl through, but some could let through a wagon. Much of the unofficial demolition seemed to be the work of mischievous youths. "Great seriousness in children's play" declared one paper, describing how schoolchildren set upon the wall with hammers and scraps of iron.[87] A couple months later, as more unofficial gates appeared, the newspaper joked that, if Berliners continued to be so industrious, by the time the official negotiations were finally complete, there would no longer be a wall to remove. Expressing the growing disjuncture between the wall's physical state and its official status as a border, the satirical periodical *Kladderadatsch* (an onomatopoeic word meaning roughly "crash") depicted the wall as an open arcade, which was nonetheless policed by constables with futile vigilance (see figure 3.3).[88]

The papers reported with evident glee how Berlin's street urchins waged battle with the stodgy and slow-witted constables who patrolled the wall. One conservative daily published a particularly jovial and detailed account of these "insurrectionists," describing how one of the young "wall-stormers" (*Stadtmauerstürmer*), upon being spotted by a constable, squeezed through a hole in the city wall of his own making. The constable charged with patrolling the border puffed toward him, gesticulating wildly in an attempt to prevent the escape. When he finally arrived at the scene of the crime, all he could do was stick his head through the small opening and yell at the fugitive on the other side. Adding insult to injury, when the constable pulled his head back through the hole, his helmet toppled out of reach into the city's "municipal area" (*Weichbild*), while he remained stuck in "wall-enclosed Berlin." The young escapee proved unwilling to hand him his helmet, so the constable was obliged to make the long trek under the hot sun to the next gate to retrieve that important symbol of his station.[89]

This story, though quite probably embellished, reveals an important theme in the public discussion of Berlin's city growth and its relationship as a growing metropolis to the Prussian state. Here, the city is cast as young, healthy, and resourceful in the figure of the urchin, while Prussian authority

3.3 Two cartoons poke fun at the futility and anachronism of Berlin's city wall on the eve of its removal. *Kladderadatsch* 17, no. 42 (September 11, 1864): 168; *Kladderadatsch* 18, no. 44 (September 24, 1865): 177, Universitätsbibliothek Heidelberg-Digitalisierung.

is personified as old, ineffective, and bound up by its own love of rules. The very overreach of bureaucratic power—attempting to control movement into and through urban space with the seemingly outdated city wall—crippled the constable's ability to exercise control even over the space within the walls. Berlin's city wall kept the official within its bounds but could do nothing against that peculiarly urban animal, the street urchin; the city had transcended the wall before it was ever removed. As in the street boy Gavroche in *Les Misérables*, it was the small figure of the street urchin that best embodied the city's spirit.

Describing the attack on the city wall in such terms was one technique for making it seem the spontaneous result of the city's own healthy and youthful exuberance. That most of those chipping away at the wall were described as young boys and adolescents recalled contemporary comparisons of the growing city itself to an adolescent and the wall to a pair of trousers grown too short.[90] The authors of these articles no doubt exaggerated the degree to which Berliners took the wall's demolition into their own hands, but they did so with a sense of humor, playing up the incongruity between the small figure of the street urchin and the largeness of the role of urban "revolutionary." But although the press made teasing use of a vocabulary of revolution to describe the surreptitious removal of stones from the wall, speaking of "attacks" and describing the attackers as "insurrectionists," this language hinted at a more serious theme as well.[91] In sidelong references to 1848, history seemed here to be repeating itself not quite as farce but as a child's game.

To some observers, the popular desire to take down the wall showed natural good sense in Berlin's citizens that was not shared by the city's administrators in their desire for an open, growing city. A playful and harmless civil disobedience, opening up unofficial city gates seemed a healthy, productive alternative to more destructive reactions to Berlin's spatial mismanagement, such as housing riots. Just a couple years before in 1863, tenants and police had violently clashed over evictions where rents were rising. Economist and housing reformer Julius Faucher saw the impatience of waiting Berliners evidenced in the wall's countless new holes, even as administrators seemed unable to discard this relic of the urban past. The city's shortage of space meant a shortage of housing, particularly affordable housing for the poor and working classes. If the wall came down, Faucher thought, the city would grow more rapidly and housing would be cheaper, more abundant, and less dense. Hence, the public desire to get rid of the wall as soon as possible must also be understood in the context of the emerging debate over urban housing reform.[92]

Another way to make the fall of the wall seem natural, inevitable, and an expression of the city's collective spirit was to obscure the actors altogether. Several accounts described the surreptitious demolition of the city wall as invisible and stealthy, even "wonderfully secret."[93] Berlin's leading financial paper reported that when an "unseen hand" opened up several new passages, locals were universally "grateful to the invisible hands that have opened up easy access from their shops to the residences on the other side of the wall."[94] Here, the metropolitan spirit was embodied in those "invisible hands," which conspired with the interests of business and capital. The article deployed the language of urban planning and improvement to describe these anonymous acts of civil disobedience. The street urchin and the upstanding businessman made unlikely allies, but here the growth of the city compared to the interest on a capital investment was enabled by those unseen and unruly Berliners who were stealthily chipping away at the wall. As in Paderborn, this was a de facto physical disintegration of the wall at the hand of urban residents themselves that eluded official schemes.

As something hidden, unseen, and of the night, the wall's removal had traces of the strange or uncanny, as if the wall were a ghost or spirit returned from a vanished past. The sound of chisels chipping away at the wall sounded to one observer like the creaking trees of the Black Forest when the Flying Dutchman passed overhead. The process by which the wall came down is again presented as the work of strange and unseen hands: "Below stand uncanny [*unheimliche*] forms who work diligently at their nightly work, sentries are placed, to warn of approaching constables . . . one stone falls, a second—finally, a joyful cry heralds the completion of the hole, though still small, and from the other side a hand stretches through and is grasped and shaken by unknown hands—a new brotherhood!"[95] Another observer described the nighttime figures chipping away at the wall as "a swarm of hot-blooded little ghosts," while the *Kladderadatsch* hoped the new holes in the wall were not the work of a dragon. The wall itself was described as unnatural, or even supernatural. It was designated an evil spell, and as a sea serpent coiling its tail around the city.[96] In many of these metaphors, the wall's nocturnal dismantlement combined the mystical and the mundane. It was a process that seemed to be simultaneously natural and supernatural, both inevitable and revolutionary.

The removal of Berlin's wall at this late date, though joyful, also seemed foolish; an embarrassment for being so delayed. Since the wall was composed of thin brick only sixteen feet tall, in a number of places around the city the many-storied rental barracks that had sprung up alongside it dwarfed it in height. The press celebrated the stealthy Berliners who sneaked out at night to take down the wall of their own accord and routinely complained of the city's slow progress in undertaking the project. The papers reported on the painstaking negotiations between city, fiscus, and the abutting landowners.[97] These stories contrasted the selfishness of competing interests on the border with the "selflessness" of those Berliners who came, even from far away, to help tear down the wall themselves. The *Börsen-Zeitung* claimed that the city's youths were forced to take on the task themselves in the face of official "obduracy" (*Halsstarrigkeit*). The *Spenersche Zeitung* observed that city officials appeared incapable of responding to even the most immediate needs of traffic and safety. "In spite of numerous applications and repeated cries of pain," the paper complained, "provisional openings have yet to be broken through even in those places where the standing wall intersects major traffic arteries."[98]

Into early 1866, as demolition progressed only slowly, Berliners continued to break new holes in the walls for convenient passage. The press largely lost interest after the first couple of months, but in February the ministerial Bau Kommission (the building commission) complained that in some places the wall was in danger of toppling "because the public is not only continuing to break through new passages—there are now twelve larger ones and numerous smaller ones—but has also begun to pull out the foundation stones, which lie exposed on the inner wall." Popular press accounts had highlighted the playful, or possibly anarchic, elements among those who secretly chipped away at the wall. As the building commission's report indicates, however, Berlin's "wall-stormers" were motivated not only by their natural desire for freedom of movement but also by the enticement of free building materials. So weakened was the structural stability of the wall that emergency funds had to be appropriated to hurry the project along. Mindful of unrest, the quickened removal was approved as a labor creation program so as to prevent an emergency situation among Berlin's working classes—quite probably the same ones who were plundering the wall for stones. Once the tempo of work was increased, a final major piece of the wall was removed in October 1866.[99]

The removal of Berlin's tax wall was one of the final steps in a process that had taken well over a century, in which an elaborate border apparatus that visibly and physically demarcated the urban edge had been dismantled. For the journalists and city boosters who celebrated the removal of the wall, a particular understanding of the city was at stake in forwarding this narrative of the wall's end as a spontaneous occurrence—the work, not of bureaucrats, but of invisible hands. For that brief moment in summer 1865, these reports represented an attempt to shape the story of the city's development, the story of how Berlin's wall fell, in the moment just before it disappeared; to take the initiative away from city and state officials and give it back to the city itself. Contemporaries construed this event—although practically of less concrete significance in the spatial development of the city—as a compact symbol of the longer transformation by which the early modern city had become a metropolis. Charles Taylor argues that it is a characteristic of modern "narrations of change, growth, development, realization of potential" that they often have "once-for-all moments: of founding, revolution, liberation."[100] In the fall of their tax wall, Berliners sought such a once-and-for-all moment encapsulating a protracted process of change and growth as a single moment—a moment of everyday and bloodless transformation. In tales of a wall that fell victim to the "metropolitan spirit," the triumphant story of urban growth combined the unconsciousness of an organic process with the explosiveness of revolutionary violence. The sense of a temporal disjunction—that the removal of the wall was an event that came "too late"—gave rise to metaphors for its removal that evoked both the unnatural and the supernatural.

The cases of Paderborn and Berlin reveal the scrambled relationships between how city walls operated both as symbols and as functional boundaries in the latter half of the nineteenth century. The story of the urban border in nineteenth-century Prussia cannot be reduced either to the inconsequential and anachronistic preservation of an outmoded era or to the repressive domination of the municipality by a state ruled with a "citadel" mentality.[101] Instead, the relationship between centralized policies that treated city walls as universalizable borders and the material reality of those walls themselves clearly demonstrates a more complex view of the relationship between the city and the state as played out in the struggle over who was to care for Prussia's many crumbling walls in the period of their growing obsolescence. The stone and brick that remained standing, a visible reminder of the city's older identity, also allowed the adaptation of the city wall as symbol to a new set of social, economic, and political tensions in the second half of the nineteenth century. In this area as in so many others, the period after 1815 was less a restoration of the old and more an awkward hybrid of the old and new.[102]

Although Berlin, a metropolis and a capital, was at the center of a growing state and Paderborn grew relatively little throughout the nineteenth century, how these two cities—both their publics and their city governments—dealt with the walls themselves had much in common. In both cities, the wall took on outsized significance after its removal but excited little interest while it still stood. In both cities, it seemed that the wall came down "too late"—an acknowledgment of the ways in which realities on the ground did not match the ideal-typical story of urban growth in the nineteenth century. Finally, in both cities the disjuncture between the physical and symbolic walls created space for conflicts over control between the city and the state. Where Berlin became really distinct was in what followed: how the space beyond the wall became a landscape not of departure but of arrival. As the center of a new German Empire, the periphery around Berlin became a place where Germans dreamed up new versions of what the empire could be and different versions of the roles different groups had within the city. It became a site for contested visions of the organization of urban space and the uses of nature.

4 The Shantytown Frontier

City Planning and Wild Settlement on Berlin's Urban Periphery

In the year 1872 two "New Berlins" were dreamed up on the frontiers of the imperial capital. One was a plan, conceived on paper by an eager housing speculator who wished to create space for the villas of the wealthy in a city rocked by working-class unrest. The other "New Berlin" appeared, spontaneous and unbidden: a collection of shacks built in desperation by workers driven out of the city amid skyrocketing rents and a catastrophic shortage of housing. The first was planned and unbuilt; the second was built but unplanned, transformed into a "New Berlin" by the words and imaginations of excited observers and participants, eager to see a hopeful future in the tumult of displacement generated by the city's growth. The parallels and contrasts between these twin visions of the German capital reborn on its frontiers reveal much about the tensions between planned and wild development, between settlement and nature, and between urban growth and class segregation in the German capital. The focus of this chapter is on the side of spontaneous growth as working-class urban activism and on the tension between what was planned and unrealized and what was uncontrolled and unexpected on the edge of the city in Berlin. Both aspects offered new ways of thinking about the urban frontier that engaged the history of the city and also inscribed visions of a future for the capital of the new German Empire in the spaces of its periphery.

One way in which Germans envisioned the edge of the city in the latter decades of the nineteenth century was through the horizons of ambitious urban planning schemes. A decade before the height of the housing crisis spurred by the boom that accompanied German unification in the early 1870s, James Hobrecht's 1862 Berlin city plan formed the foundation for a new national discourse around German planning practices. Both the spontaneous working-class shantytowns and the planned residential developments of real estate developers such as Johann Anton Wilhelm von Carstenn and Johannes Heinrich Quistorp challenged these emerging city planning practices.[1] Before the 1860s, there were only a few available models for ambitious municipal planning. In Vienna, the city used the land freed up by the removal of its expansive fortifications to construct a monumental ring road (*Ringstrasse*), accommodating and glorifying urban professionals in its aesthetics and infrastructure. In Hamburg, not a defortification but a catastrophic fire in 1842 provided the circumstances for an ambitious rebuilding plan in what Fritz Schumacher called the best example of the "art of overcoming the big city through urban planning." After the 1860s change accelerated as a new class of technocrats and reformers pursued ambitious projects of restoring order to the growing city and sought to reconcile public and private needs, aesthetics and pragmatics, as Brian Ladd explores in his study of German city planning.[2]

Histories of nineteenth-century city planning focus largely on the role of developers (private and public), municipal administrators, and the urban professional classes they served.[3] On the other hand, there is also a rich historiography of how the "other half" lived in the growing industrial city. In his study of cholera and planning in Hamburg, Richard Evans describes in stomach-turning detail the ways in which the environments of urban industrial capitalism shaped the lives, health, and opportunities of the urban working classes, whose access to adequate housing, clean water, and unadulterated food were all challenged by the modern city. There is also a vibrant history of studies of the Berlin underworld, going back to Hans Ostwald and Magnus Hirschfeld around the turn of the nineteenth century, describing working-class culture and the intersections between policing and the urban underworld.[4] There has been much less consideration of the spaces created by the urban working class themselves. Between the schemes of urban planners and the plight of the urban poor is another history in which those populations usually left out of the city-building narrative shaped their spaces through use, construction, and reappropriation.

In the "New Berlins" of 1872, working-class residents were not simply victims of the urban environment but, rather, active shapers of it, both alongside and sometimes intersecting with more traditional planning discourses. The shantytown and other informal unplanned urban spaces can be understood as part of a process of radical performance and space making on the part of Berlin's working class within a broader history of makeshift construction and provisional spaces in the city. Other recent scholars of the history of squatting have similarly shifted from treating working-class housing as a mere reaction to circumstances to seeing in it the expression of ideas about dwelling and claims to spatial power within the city. This is what Lisa Goff has called the "working-poor ideology of dwelling" and what Alexander Vasudevan calls a "spatial grammar of squatting."[5] Berlin's shantytowns take their place among other instances of worker self-help in the urban environment. Their construction by working-class Berliners represents a kind of "settler urbanism" that transformed the edge of the city into a frontier—a space of spontaneous development, settler independence, and mixed economies and environments that challenged the clear division of city and countryside.

Although self-built housing played a relatively small role in German city making, the self-help movement in other forms played a significant role in workers' organization and the making of urban spaces.[6] Allotment gardening was among many reform efforts that straddled the boundaries between bourgeois reform and worker self-help. The allotment gardening movement linked concepts of home and nature to the idea of the city. Boundaries between allotment garden settlements and shantytowns in Berlin were porous and contested, and these transitional spaces became characteristic features of the capital city's urban peripheral landscape in the twentieth century. Like the shantytowns of the founders' era (*Gründerzeit*), working-class uses of nature such as wild garden settlements and gleaning practices rendered the transitional space of the urban edge into a landscape of precarity visible to artists and the public. These models of improvisational use link nineteenth-century Berlin to a global landscape of development and highlight surprising continuities between the industrial and postindustrial city, offering a more democratic view of the city and reframing squatting as an act of political resistance with continuing significance today.

CITY PLANNING, CLASS, AND THE POLITICS OF HOME

The challenges faced by the Prussian city that became the capital of the German Empire in 1871 were in many ways anomalous, although Berlin dominated the national discussion of urban development like no other German city.[7] By the end of the nineteenth century, German cities enjoyed a global reputation for excellence.[8] Their order, architecture, and infrastructure combined the advantages of modern urban life with the preservation of historical heritage and visual interest. These internationally admired accomplishments rested in part on a tradition of municipal autonomy and strong local governance that allowed city authorities to shape urban spaces effectively—advantages Berlin often did not share. Those cities that received the most international praise, such as Frankfurt am Main, were also those where municipal independence and urban liberalism were the strongest.[9]

The development of modern urban planning practices in German cities was not, however, simply a product of historically rooted patterns of urban governance. German urban spaces reflected the desires, needs, and worldview of urban elites that coalesced toward the end of the century into a distinctive variety of German "bourgeois modernism," of which one of the most salient features, as Maiken Umbach has defined it, is its integral relationship with the material culture that formed its most significant legacy. Through the landscape of the German city itself—the urban spaces, from streetscapes to city parks, suburban villas to historicist monuments—the material form of the city became for the urban professional classes a technology of governance and control, through which the power of liberal urban elites was cemented.[10] Unplanned "wild" urban developments such as shantytowns and allotment gardening challenged both urban planning initiatives and the bourgeois modern ideas behind them. At the same time, the bourgeois politics of urban homemaking clearly shaped the overall perception of improvised settlements, which observers often described as mimicking planned urban developments and middle-class ideals of home, family, and even recreation.

The close relationship between German bourgeois modernism and material culture was defined by a distinctive emphasis on the home and the link between domesticity and nation. Housing and characterizations of home played a key role in the development of urban planning and reform movements, whose origins can be traced back to the close association of homemaking and local place with national identity through the multivalent German concept of the Heimat. Historians have recently detailed the many ways in which this emphasis on home shaped German practices of place making—from the role played by the domestic arts and the household in German national identity, both at home and abroad, to the strong link forged between cultural and natural landscapes in the homeland preservation (*Heimatschutz*) movement, and the profound importance of land and the concept of home animating many of the diverse reform movements of the generation of 1890.[11]

Ambitious modern city-planning impulses emerged as part of a slate of late nineteenth-century projects of bourgeois reform on both sides of the Atlantic. One of the primary purposes of modern city planning was to configure the city as a space for homemaking on both the macro scale of the city as a whole and on the micro scale of the individual residence. The link between city making and homemaking was made explicit in the language used by city reformers and planners to describe the desired spatial order of the city. One German observer of Haussmann's Paris linked city planning to housekeeping by comparing slum clearance to sweeping the city as if with a "large broom."[12] The statistician Ernst Bruch in an 1870 critique of Berlin's historical development and spatial form evoked the words of a city planner arguing that the ideal city ought to function like a home. This meant first and foremost that city space should be clearly ordered in a way that was defined by its daily uses. No one should have to ask in a city where to find the shops, factories, or the residential blocks, any more than a new visitor to a house would have to ask which room was the kitchen. Berlin's spatial disorder, stemming from its historical development, rendered it not just ugly and impractical but also unhomely.[13] Two decades later, the planner and architect Josef Stübben drew the same parallel between the well-ordered city and the home in his praise for a Haussmannized city: "It is just as easy to find one's way about as in a clearly designed house. This gives us a feeling of security and pleasantness . . . that the visitor in the unsystematically designed city will always miss."[14] The disorderly city was unhomelike in its organization and also *unheimlich* (eerie) in its mood.

Although such expressions on the part of bourgeois reformers linked urban planning to the politics of the home on the level of spatial allegory, the association was more than just symbolic. The Haussmannization that had rendered Paris so homelike to Stübben and other reformers entailed the clearing out of working-class quarters and the expulsion of impoverished dwellings to the periphery.[15] The perception of urban disorder on the part of city planners implicated contemporaneous conversations about economic class within the city. Since at least the 1830s, middle-class observers had found working-class urban dwellings to be insufficiently "homelike" in their crowding, unhygienic conditions, and perceived impermanence. Friedrich Sass wrote of the large housing complexes outside Berlin's Hamburg Gate in 1846 that this kind of degraded abode was itself barely distinguishable from homelessness. Thirty years later, Reinhard Baumeister wrote that a temporary residence in a rental barracks resembled nothing more than the "stone tent" of a modern nomadic life. Concerns about dwelling instability were particularly high in Berlin. In the first volume of the city's statistical yearbook in 1874, the statistician Hermann Schwabe penned a detailed report on the "nomadism" of

Berlin poor and working-class populations, in which he both quantified and visualized the relationship between levels of rent and frequency of moves: the lower the rent and the poorer the housing, the more often Berliners moved. In the cheapest category of apartments, almost half of all residents moved each year.[16]

The goal of city planning and housing reform was to make the city an environment conducive to middle-class homemaking. Working-class housing challenged this project both in the characteristics of working-class urban spaces and also in the precarity of working-class lives. Freedom of movement challenged the status of the city as a homelike place, linking conversations over mobility and emigration to urban housing reform. In concern for both the internal and external features of lower-class housing quarters, the creation of appropriate living space was at the center of the projects of social and religious reformers, planners, and municipal governments as they confronted the problem of urban poverty.[17] At the same time, the German city was also home to a large and growing working class with its own alternative culture, social milieu, political organizations, and modes of understanding and using urban space.

Working-class neighborhoods tended to develop in two main types. First, there were dense inner-city neighborhoods. Later, as density increased, downtowns became more commercially valuable real estate, and industry developed outside the city, working-class neighborhoods shifted to peripheral neighborhoods. These neighborhoods were more segregated by class and were characterized by large housing complexes. Housing in these neighborhoods included employer-built settlements and were often located near the factories, which also moved into the urban periphery over the second half of the nineteenth century. In Berlin, tall apartment buildings clustered outside the city gates.[18]

Nineteenth-century commentators, in part by observing the clustering of housing and institutions along major roads out of the city, thought of the urbanization process as proceeding from the center: the city spread or grew outward. Spatial and economic studies of urban development make clear the inadequacy of this model, but it has continued to structure many assumptions about the history of urban space, particularly in the realms of social, cultural, and planning history. Instead, as Ingrid Thienel has demonstrated in the case of Berlin, urban growth proceeded as much through the independent urbanization and industrialization of extra-urban regions. The enclosure of common land because of industrialization and the restructuring of adjacent communities outside the city made possible the creation of an urban peripheral region that was later incorporated into the city. The edge of the city was shaped by the interaction between a growing city and an urbanizing periphery. Workers' housing clustered along the city border because of this interaction. For example, an area especially desired by industrial workers was the Wollankstraße, which ran between Schönhauser and Rosenthaler Gates, because it was a reasonable distance from the factories in Oranienburg and directly connected to the Rosenthal suburb as well.[19]

If one shifts perspective away from seeing the urban periphery as an incompletely urbanized hinterland in the process of an urban colonization process, then one can see that where workers lived to some extent resulted from strategic choices and not simply from the overflow of a crowded city. These neighborhoods were also sites of innovative projects of working-class homemaking and place making, even if middle-class and professional observers were unable or unwilling to recognize them as such. These projects become more visible when one looks not only at the individual residence but also at the features of neighborhoods as a whole, taking a different approach to the relationship between public and private space. Because working-class urban residents had very high mobility, even moving several times in a single year, homemaking did not focus on the individual unit. As Adelheid von Saldern points out, workers who relocated within the city usually remained within a relatively small and well-defined region. For this reason, they formed attachments not to individual dwellings but rather to blocks or neighborhoods.[20]

These "workers' quarters" fostered institutions, customs, and dense communication networks that developed into a distinctive urban "culture of poverty" and provided a sense of identity and belonging that bound residents to their neighbors and their physical environments. This culture was closely linked to the spaces and spatial practices of the quarter. Where apartments were small and crowded, shared and public or semipublic space such as communal kitchens provided valuable extra space. The infamously dark and grimy courtyards of Berlin's rental barracks housed communal self-help projects, serving as the distribution sites for workers' retail cooperatives. They provided spaces for communal work and socializing and had their own distinctive sounds and entertainments, such as the "courtyard music" (*Hinterhofmusik*) of wandering musicians, with their harps, violins, and accordions. Children played—and were tended to—in the streets. In the strong attachment of workers to their ostensibly disreputable addresses one might even discern a kind of conscious defiance or passive resistance to the spatial order and reforming spirit of the bourgeois city.[21]

The cartoons and illustrations of Heinrich Zille, the most prolific documenter of imperial Berlin's social milieu, show how working-class "making-do" shaded into defiance, independence, and distinctive place making. His images document the ways in which the urban working classes occupied both private and public spaces and blurred the distinction between the two. As Amanda Brian has argued, Zille's drawings and photographs in the working-class quarters and far-flung settlements of Berlin reveal a two-way conversation between bourgeois and proletarian modes of life in the city. For example, in Zille's drawing "Im Bürgerheim" (In the home of a burgher), a working-class family makes itself at home in a neighborhood pub. Through their presence and use of the pub, the family transforms a public space into a private and homelike retreat—a baby cries and bounces on his or her mother's knee, while an older daughter takes a nap with her head on the table.[22] The background is that of a pub and yet contains familiar signifiers of a German household. Advertisements for beer decorate the walls alongside placards that appear at first to mimic the framed sayings that typically decorated domestic walls. Brian argues that Zille's drawings and photographs in the working-class quarters and far-flung settlements of Berlin embody a "proletarian modernism," the humor of which often stemmed from the translation of proletarian spaces and customs into the culture of bourgeois modernism—a kind of humor that, as we shall see, was also evident in the self-conscious staging of Berlin's shantytowns in 1872.

The later nineteenth-century city, then, was the location of both bourgeois and working-class homemaking and place-making projects. The tensions between these spaces produced two related conflicts over urban planning and class: how legible class differences should be in urban space and how segregated different classes should be from each other. Questions of both the legibility and the segregation of economic classes in city planning crystalized around the debate over the implementation of James Hobrecht's plan for Berlin in the 1860s. Responding to the challenges of rapid urban growth and the shifting relations between working-class and middle-class spaces in the city, Hobrecht's 1862 plan was one of the earliest urban plans to both encompass the city as a whole and to anticipate and provide for future city growth. His vision for Berlin gave the city the fundamental form and character it still has today. Subsequent controversies over the plan's virtues and failings also initiated a new German urban planning discourse in the following decades.[23] Even if Berlin was not representative of German cities, its planning challenges decisively shaped the national dialogue on urban issues.

The Hobrecht plan was long seen as epitomizing the failings of nineteenth-century city planning. While real estate speculators and developers saw the plan's strictures as dangerous impositions on private enterprise and free growth, social reformers concluded that Hobrecht did not do nearly enough to regulate growth and protect the health and interests of urban residents, especially the most vulnerable, from unhygienic conditions and unscrupulous housing speculators. Often compared to the restructuring of Paris under Napoleon III and Baron Haussmann, Hobrecht's plan was far from a whole scale "Haussmannization" of Berlin either in its original intent or in its incomplete execution. Hobrecht strove to avoid the unnecessary expropriation of private property for a combination of practical and legal reasons.[24] The primary criticisms concerned the excessive width of the streets and size of the blocks, which together allowed for the monotonous windswept streetscapes and large housing complexes for which Berlin is known. Critics blamed Hobrecht's lack of foresight, failing to account for real estate speculation and for his excessive ambition, attempting to control the future by extending its rules far beyond the city's built periphery in a so-called one-hundred-year building plan. These criticisms found an early and influential voice in the writings of Berlin statistician Ernst Bruch, who penned several critiques of Hobrecht's plan around the year 1870. The nadir of Hobrecht's reputation came with city planner Werner Hegemann's widely read excoriation of Berlin as Germany's capital city in *Das steinerne Berlin* (*The Stone Berlin*, 1930).[25]

Reevaluations of Hobrecht since the middle of the twentieth century have argued that his plan's very limitations evidence a possible prudent humility: leaving room for adaptation, avoiding expropriation and demolition wherever possible. Urban historians like Claus Bernet have recognized that Hobrecht cannot be blamed for the unexpected ways in which his plan was later adapted and implemented. Hobrecht did not plan green spaces directly, but he intended the large blocks and many squares of his plan to allow for the insertion of ample green space. Sometimes this worked as intended: wide streets allowed for treed boulevards, for example. In other places, the plan was overtaken by speculation that squeezed out room for green. In its implementation, planners gave traffic flow priority and many of the open squares that Hobrecht intended were never built.[26]

Why, then, was there near universal condemnation from the 1870s through the 1930s? The reaction against Hobrecht's plan can be better understood if placed in the context of larger controversies about politics and class emerging in the 1870s. These debates also set the stage for understanding later conversations about unplanned settlements as a reaction against the Hobrecht plan. The first of these controversies concerned the question of the centralization of power in the Prussian capital and the proud German tradition of urban independence and regional diversity. Ernst Bruch's 1870 book on Berlin's development begins with the observation that since national growth had increasingly become synonymous with urban growth, the moral and hygienic questions of city development had become questions at the heart of the national character. Bruch feared that the empire's over-centralization concentrated too much on Berlin and would direct resources and attention away from Germany's wealth of diverse and dispersed urban centers. Bruch feared that "true Berliners" had begun to think of their city as "the city" and "the entire remainder of the state as its province."[27] In this context, denunciations of Hobrecht's "one-hundred-year building plan" implied a critique of an overreach of power from the capital with implications not just for the immediate hinterland but for the whole empire.

A second controversy over the Hobrecht plan involved the spatial separation and legibility of economic classes in the city. Bruch criticized Hobrecht's Berlin for an oppressive sameness, which rendered the city unhomelike and streets illegible in terms of both their function and the class of their residential inhabitants. Ostensibly an aesthetic concern, the accusation of monotony hid tensions over class relationships in the city. Bruch wished for a city that would be "un-mixed," in which the visual appearance of a street reflected its place in the social and economic order of the city.[28] Less socially conservative city planners such as Stübben and Baumeister evinced more ambivalence about the economic segregation of the city, but they too tried to make social class legible in the urban environment. Hobrecht's plan went against this impulse. One of its most characteristic features was Hobrecht's dedication to economically integrated neighborhoods. In 1868 he argued that the correct approach to the poorer classes of the city was "not 'isolation' [*Abschließung*] but 'permeation' [*Durchdringung*]."[29]

Hobrecht believed that dense, integrated housing created healthy cross-class community and grassroots charity in daily encounters in the streets, shared stoops, and small acts of neighborly generosity. To some extent this worked. In the conflict between Hobrecht and Bruch's approach to segregation and zoning in the city, the 1860s and 1870s marked a key moment in the transition to a modern urban paradigm in which class became less legible from building exteriors, a model that reached its German apotheosis in the apartment blocks of the Weimar era and later twentieth century.[30]

Finally, the Hobrecht plan debate continued an older conversation about what kinds of housing could be conducive to the creation of home and domesticity. Earlier housing reformers had often advocated for detached single-family homes on the English model as the gold standard of housing for the working classes. By the 1870s many housing reformers acknowledged that this model was unreachable in dense, rapidly growing cities, even as the idealization persisted of housing based on detached single-family dwellings instead of multilevel apartment complexes.[31] The orderliness of the city, class segregation, and the type of housing conducive to the creation of urban homes were all linked themes in the debates over urban planning and the nature of growing industrial urban spaces in the second half of the nineteenth century.

Contemporary observers interpreted the housing crisis of 1872 caused by the city's rapid growth and subsequent shortage of affordable apartments largely in the terms set by the Hobrecht plan and city-planning discourse of the previous decade. The reaction focused on the segregation of classes within the city and on the visibility of urban poverty, on the choice between single-family housing and apartment blocks, on the conflict between centrally planned and market-directed urban growth, and on the relationship between form and function in urban planning—whether the built urban landscape ought to reveal its history of development or appear as a ready-made whole. Furthermore, observers placed this conflict in the context of an emerging global landscape of makeshift housing. Some observers celebrated the freedom and entrepreneurship of the unplanned shanty settlements; others saw them in the context of a housing crisis expressed as extreme and unsustainable segregation of residential neighborhoods by class. Lower density housing may have mimicked middle-class housing ideals, but it also entailed greater separation of classes. Saldern identifies a movement toward urban anonymity and illegibility of class in city space in the strong reaction against the shantytowns. That is, one problem with the shantytowns was that they rendered urban poverty separate and visible to all as a spectacle of city building on the urban horizon.[32]

Working-class neighborhoods offered alternative visions of home in the city. Historians of the city have not done enough to integrate this culture into our understanding of urban form. One reason for this omission is that it often remained hidden and hard to document. In the shantytowns, however, working-class constructions of home came into view, as they also did around the turn of the century in the art of "proletarian modernism" such as Heinrich Zille. The shantytowns offered another vision of a "working-poor ideology of dwelling" in the city—unplanned and spontaneous in its built structures, not just its communal forms. While working-class homemaking in the city was often hidden or only became visible when it was the object of social regulation and policing, the very peripheral position of the Gründerzeit shantytowns just beyond the recently removed city gates placed this project of working-class homemaking at the very center of the public view of the growing city. Just as the shantytown brought the process of urban expansion and its inequalities out into the open, it could also be seen as a masquerade—a confusing and ambivalent landscape of mimicry obscuring the boundary between the inventiveness born of extreme want and middle-class striving.

SHANTYTOWNS AND THE "WORKING-POOR IDEOLOGY OF DWELLING"

The economic boom that accompanied the founding of the German Empire in 1871 led to an unprecedented housing crisis in the capital.[33] The population was in extraordinary flux. In 1871, roughly 133,000 people moved into the city and 78,000 moved out. In a city with a population of 800,000, there were 55,000 new residents who had come to stay. By April 1872 there were 15,000 people without homes. Riots broke out in working-class neighborhoods. Whole families squatted curbside with their furniture.[34] Some of the evicted sought refuge in train stations, others in overturned rowboats and abandoned train cars. Some even resorted to digging out shelters in sandy soil at the city's edge.[35] The most visible consequence of the crisis was the appearance of shantytowns just beyond the old Halle and Kottbusser Gates, to the south of the city in present-day Kreuzberg, and outside the Frankfurter and Landsberger Gates to the east. Because the city had spread less quickly in this direction in the recent decades of rapid growth, here the edge of settlement still mostly coincided with the recently removed city wall. On vacant land largely owned by the city, the homeless constructed huts from scrap wood, straw, and old furniture.[36] In May 1872 local papers reported that the largest of these encampments already housed ninety families in fifty-two improvised dwellings.[37] According to some estimates, the population in this settlement alone would approach two thousand before it was cleared in September. As they grew over the spring and early summer, the shantytowns took on the features of more permanent settlements. Some of the huts boasted gardens, fences, painted walls, windows, and flowerboxes. Stores and workshops appeared amid the residences. The self-built shacks were arranged along dirt paths serving as streets, some with their own street signs.[38] The squatters appeared in the press as urban pioneers, their self-made dwellings progressing from haphazard shacks to enviable homesteads through their own labor.

In spite of these trappings of permanence, mimicking the formal process of the city's planned expansions in these years of whirlwind growth, the shanties were destined for rapid demolition. As concerns about urban working-class radicalism rose, the city cracked down on the squatters. In the name of fire safety and hygiene, city officials cleared the shantytowns before the end of the summer. City officials saw the shantytowns as an expression of dangerous working-class political impulses. In order to ensnare workers with socialist leanings, police patrolled the crowds gathered to protest the squatters' evictions. Although housing shortages remained a serious problem in Berlin, the worst of the crisis eased by the following summer, as the boom of the empire's founding years went bust.[39]

However short-lived the shantytowns of 1872 form an important chapter in the history of Berlin's urban development. During their brief existence they became the object of fascination in Berlin and across the nation. Artists documented their appearance for national publications such as *Die Gartenlaube*, *Die Illustrirte Zeitung*, and *Über Land und Meer*, printed alongside eyewitness reports of the "new city district" in Berlin (see figure 4.1).[40] Local newspapers tracked their growth over the summer. Correspondents in Frankfurt and Hamburg eagerly covered the events, even as their home cities suffered housing crises of their own. The unplanned settlements even garnered attention on the other side of the Atlantic, where reporters viewed them through the lens of German transatlantic migration. Encouraged by press reports, curious onlookers ventured out to view the shantytowns for themselves.[41]

What observers saw in the shantytowns reflected their preconceptions and preoccupations about the nature of the modern industrial city. Some onlookers saw a return to barbarism and evidence of the degrading effects of the modern urban landscape on the poor who lived there; others saw a landscape of hope: of workers' resourcefulness and powers of self-help, of rugged individualism and a spirit of adventure, even a welcome return to nature in the face of urban squalor. In these accounts, the makeshift settlements appeared as a frontier landscape, sharing in the spirit of the frontier towns of the American West, but also as a tiny self-made version of the nature-nestled villa settlements of Berlin's own West.

As one landscape of urban transformation Berlin's Gründerzeit shantytowns have been inscribed with competing meanings by retrospective commentators just as they were by contemporary observers. They have been claimed as conclusive evidence for the bankruptcy of nineteenth-century urban planning models, as motivation for Berlin's commitment to the high-rise "rental barracks" (*Mietskaserne*) model of worker housing, as an early precedent for the squats of the twentieth century and testimony to Berlin's lasting status as a center of radical housing politics, and as an origin point for the Berlin wing of the German allotment gardening movement.[42] Although the 1872 shantytowns must be understood in the context of a longer history of

4.1 An illustration of the Kottbusser shantytown from the *Illustrirte Zeitung* in June 1872 depicts the homeless settlement as a space of hardship but also a space of resourcefulness and pride. Knut Ekwall, *Barackencolonie obdachloser Familien am Kottbuser Damm bei Berlin*, 1872, Stiftung Stadtmuseum Berlin.

informal settlement on the urban edge, they also stand out as a discrete event and media spectacle. At no other time and in no other place did impermanent and self-built settlements capture national attention in the same way.

The contemporary celebration of Berlin's peripheral shantytowns in the popular press of the time resonated with the growing reaction against both comprehensive planning efforts and speculative real estate development. The shantytowns preserved their place as evidence of the urban housing market's most spectacular failings in housing reform literature into the twentieth century, and even in the squatters' movement of the 1970s and 1980s. Many of those figures whose depictions of imperial Berlin's urban housing crises have remained most influential arrived in the city in the prior years and lived in the vicinity of the shantytowns including housing reformer Adolf Damaschke and artist Heinrich Zille.[43]

The urban poor and working classes created urban spaces that sometimes overlapped with and reinforced the projects and goals of urban planners and reformers and in other ways resisted and overturned them. This has been especially true in the largely ignored realm of makeshift and self-built housing.

If we take the form of the shantytown not simply as the product of extreme want and desperation but as an architectural form and created object, the shanty itself has a particular history and geography that deserves attention in the context of urban expansion planning and, particularly, on the peripheral spaces of cities.

In the 1860s and 1870s, the shanty was an architectural form associated with the frontier, and particularly the North American frontier. Recent history of the nineteenth-century American landscape has begun to take more seriously the history of the shantytown as an architectural and communal form. Lisa Goff argues that the forgotten landscapes of urban shantytowns "constitute an alternative vision of American urban space . . . that embodies the working class's evolving vision of itself between 1820 and 1940." Goff argues that the shanty represented a specifically "working-poor ideology of dwelling." This ideology resisted the "republican spatial imagination" of the checkerboard grid to follow an apparently chaotic structure more akin to that of the village or small town—clustered around central points and following the natural features of the landscape. The shantytown rejected the separation of spaces of commerce, production, residence, and agriculture, as well as a clear separation between private and public spaces, favoring an amalgam of spaces following the needs and daily habits of residents rather than the logic of a city plan.[44]

As a nineteenth-century urban form, the shanty was not an American peculiarity. Rather, a particular kind of self-built housing functioned as working-class resistance to control, oversight, and planning in cities on both sides of the Atlantic. Goff draws on the work of Ann-Louise Shapiro on working-class housing in Paris to suggest that, by creating their own urban neighborhoods "off the grid," working-poor residents of Brooklyn and New York demonstrated their resistance to forms of dwelling forced on them in urban space, especially the large multiunit apartment building. In the case of Paris, Shapiro argues that working-class residents resisted large housing complexes because they "represented diminished personal autonomy." They too preferred self-built makeshift dwellings constructed on the outskirts of the city. As Vasudevan describes it in his history of squatting in Berlin, squatters, like other working-poor residents of the city, engaged in the "making of an alternative urban imagination."[45]

The resonances between the work of Shapiro, Goff, and Vasudevan in very different contexts suggest unexplored avenues for comparative work on the makeshift dwelling of the working poor in Europe and North America in the late nineteenth century. Lending these resonances greater salience is the connection of European and North American cities through migration as an experience both concrete and imagined. The frequent comparisons to the American frontier and the collective experience of migration in the reactions to the shantytown developments of the Gründerzeit raise the question as to whether they exhibited a similar rejection of the top-down spatial organization of the city and the creation of an alternative spatial imaginary. This spatial imaginary is best understood in the context of not just local urban issues but also global landscapes and narratives of development.

The types of homelessness and homemaking on display in the housing crisis of 1872 were of new and unfamiliar types that challenged observers' ideas about how the residents of these shantytowns related to the space of the city. Earlier in the century, the most familiar type of homelessness was that of the vagrant or vagabond. Single, male, unemployed, and unconnected, he was a figure entirely external to sedentary, productive society, and he was associated not primarily with the urban landscape but with the open country road. In the public eye, vagabondage was related to a self-chosen "radical homelessness"; that is, a rejection of the social demand that one keep a fixed abode and stable employment in favor of a kind of radical freedom from expectations and social obligations.[46] Single male vagrants from outside Berlin made up the largest group of occupants in Berlin's institutions that provided shelter for the homeless, and women who sought shelter with the city were also described as transient and unattached. Literary documenters of the culture of urban poverty like Hans Ostwald carried this older type into the twentieth century with powerful textual representations of their own experiments in cross-class drag.[47]

In contrast, the mass homelessness that so visibly affected working-class families was a relatively new phenomenon around the time of the shantytowns. In the previous years the presence of whole families in homeless shelters elicited special disgust and concern. In 1868 journalist Gustav Rasch noted with shock that families including women and children were forced to take shelter alongside "beggars and vagabonds of every sort and every age."[48] Indignation that newly homeless families were being treated to the same conditions as vagrants and prostitutes helped to instigate riots at a Berlin asylum in 1871. In reaction, the city added separate family accommodations in 1873. In contrast to the predominantly male and unattached world of the homeless shelter, conversations about lower-class housing concerned primarily the poor quality of available housing and the transient residences of marginally housed families, for whom homelessness routinely peaked on the quarterly dates when leases came of term.[49]

The housing shortages of the 1870s brought together these two discourses on lower-class housing and homelessness, one focused on homelessness as either a choice or the result of unforced choices and the other focused on the plight of families victimized by unfair housing markets. As depicted in the popular press, the families of the shantytown were themselves practicing a kind of radical homelessness, claiming space beyond the city edge and putting their homelessness on display as a positive choice for a particular kind of life rather than allowing themselves to be at the mercy of an irrational housing market. But the form this protest took was a radical assertion of the most traditional—and traditionally "German"—form of family life as detailed in discourses on housing and home: the productive, employed nuclear family living in a single-family house, rather than the insecure and migratory life of the tenement dweller.

Popular press depictions of Berlin's shantytowns as frontier settlements found in them proof of a populist rejection of both the plagues of city living and the plans of urban reformers. Like the Berliners who dismantled the city wall in the previous decade, the shantytown residents expressed a natural good sense and understanding of urban space that thwarted urban administrators' mismanagement. In the pages of popular journals and newspapers, frequent appeals to the American West reinforced this narrative. The *Illustrirte Zeitung* referred to one shantytown as a "little America." *Die Gartenlaube* compared its residents to "American frontiersmen." An article in the *Germania* depicted the shantytown as a German San Francisco, evoking Friedrich Gerstäcker's novels of the gold rush.[50] According to this story, Berlin's working poor had been compelled by the hard conditions created by the housing crisis to strike out into the wilderness at the city's edge to make a life for themselves: the creation of an urban shantytown frontier represented a radical choice to shape the landscape and the conditions of one's own homelessness.

Beyond explicit evocations of the Wild West, the frontier myth constructed by accounts of the 1872 shantytowns shared a number of characteristics familiar from global frontier myths. For one, these accounts argued that the frontier shantytowns produced a culture of self-dependence and democratic institutions. People in the shantytowns were peaceful, hardworking, and governed by consent. Second, the makeshift settlements served as a safety valve for metropolitan tensions—the shanty dwellers' peacefulness and conservatism contrasted with the violence of the eviction riots that also plagued the city in summer 1872. The shantytowns were also "a gate of escape from the bondage of the past."[51] Having escaped the city's exploitative real estate market, the shanty dwellers were "as little dissatisfied as the American backwoodsman, who would certainly not want to exchange his fine though rough, comfortable, and healthy log cabin with our miserable tenements." In their self-built huts, they were certainly happier than they had been in their "woe-begone attic chambers and damp basement holes."[52] In order to make this a story of escape from the city, accounts often exaggerated their spatial marginality and isolation. Although they were within the new post-1861 boundaries of the city and the wall had been fully dismantled by this point, reports emphasized that the shacks were *outside* the city—beyond its gates—and exaggerated the emptiness of the surrounding landscape. Reporters took advantage of the nearby location of the old gates to narrate a visit to the shantytown as an exit from the city into a different world.

To some the appearance of the shantytowns even indicated a healthy instinct among the German working classes to return "to the land" from the unhealthy city. The *Illustrirte Zeitung* spoke of the shanty dwellers who lived in a freely chosen "state of nature" (*Naturzustand*). A speech given to the Berlin craftsmen's union about the shantytowns celebrated the instinctive good sense that Berlin's lower classes displayed by this apparent return to nature, resulting in a pattern of settlement that was more natural, and more German, than that to be found in the city, precisely for being born from necessity. The shanty dwellers epitomized uniquely German virtues in their decency, their contentedness, and their healthy "sense for nature" (*Natursinn*).[53]

Descriptions of the shantytowns (even some of the less sympathetic) emphasize their surprising orderliness in spite of the rough conditions, detailing the care that inhabitants took to make hut into home. According to these accounts, the squatters decorated the bare wood of their walls with bright paint and their windows with flower boxes of gardenias and planted vegetable gardens outside their huts. The shantytown could even appear to be a cozy "idyll." Although in fact (as the numbers of dwellings and residents demonstrated) multiple families shared shacks, accounts played up the image of the detached residence with clear and respected property boundaries.[54] The more celebratory accounts implied that families built their own dwellings, emphasizing their self-sufficiency, while other papers reported a thriving local business of shantytown carpenters.[55] These images echoed the earlier housing reform movement in its focus on the ideal of the single-family cottage, drawing on both English models and a supposed Germanic tradition, though the focus on the single-family dwelling had become marginalized within housing discourse and had come to seem like a much more radical—and problematic—solution to urban housing ills by the 1870s.[56]

Accounts habitually emphasized that the residents were poor, but not destitute. Unlike earlier depictions of extramural and peripheral slums, descriptions of the shantytowns rarely mentioned anyone who worked in a factory. The articles instead emphasized the spontaneous organization of "commerce and industry in its simplest forms" in the settlements.[57] Although Berlin's urban edge was increasingly an industrial landscape (industry thrived in the post-unification boom and, as conditions grew cramped in the city, relocated more and more to the city's edge), the shantytowns depicted in the press were dominated by small-town trade and handicraft, populated by potters, shoemakers, tailors, and other small-scale craftsmen who set up shops in their self-constructed shacks.[58]

Finally, the shantytowns were depicted as having escaped the ineffective and overbearing administration of the city, in a rejection of both the overreach represented by the Hobrecht plan and the meddlesome governance of city and police. The shantytown dwellers provided services for themselves with more efficiency than the city government. They constructed their own streets and built communal utilities, such as stoves and fountains. They provided their own security, catching and expelling criminals. The intervention of city authorities, even when well intentioned, was a regrettable intrusion on a self-sufficient settlement. The *Illustrirte Zeitung* lamented that unwelcome "ambassadors of civilization" from the city would soon "make an end of this free community."[59] In a story that was widely reprinted, a reporter for the *Hamburger Correspondent* conveyed this spirit in a quotation from an apparently relatively well-off member of the shantytown community: "'What can you do,' said a young man in white shirt sleeves, dressed in clean, fine clothing, there are no apartments for us, so you have to build one for yourself."[60]

These elements of independence and spontaneous self-organization resulted in a number of imaginings of the shantytown as its own city or even as an independent republic. It was referred to as "Little America" or "New Berlin" and given fanciful names such as "Republic Barackia" (derived from the German word for shanty, *Baracke*). The *Kladderadatsch* printed an imagined page from the shantytown's own independent newspaper, called *Die Baracke: Organ für Neu-Berlin* (The huts, an organ for new Berlin), in which news from the city center was reported under the heading "from the provinces."[61] Although the shantytowns' "genius for self-government" reminded one observer of 1848 precedents, accusations of dangerous radicalism were rare in such press accounts.[62]

While romanticization of poverty on the part of the bourgeois press is certainly part of the story, there are several clues in the sources that the squatters themselves sought to shape the public narrative of their plight, and to fashion their makeshift settlement as a form of peaceful resistance. Squatters were clearly aware of their audience and took advantage of the public interest by placing benches for convenient viewing under the hot sun and selling beer to thirsty slum tourists.[63] The press coverage—especially outside of Berlin—emphasized the ways in which the settlers exhibited the characteristic sardonic sense of humor for which the capital was known. A journalist from Hamburg observed: "The irrepressible Berlin sense of humor can be found in these quarters as well. Some of the huts bear high-sounding names written in chalk, like Hotel Henkel. The paths by which one reaches the individual huts are called private streets and the individual huts are constructed on top of artificial hills covered with lawns [*Rasen*], so that they resemble castles."[64] Also contributing to the castle-like appearance of the huts was perhaps the practice of digging out a trench and building a "protective wall" with the dirt around the walls—presumably to protect the huts from flooding. The same Hamburg correspondent evinced surprise at the openness and readiness with which the residents of the shantytown answered questions about their plight, taking this lack of apparent shame as evidence of their satisfaction and pride.[65]

The difficulty of interpreting these attributions of "Berlin humor" to the shanty dwellers in the national press is that it is nearly impossible to tell what of it comes from the observed and what from the observer. The residents of that shantytown may have used chalk and street signs to frame their squatting as resistance to the conditions of Berlin's urban space, but it was the reporters who compared their huts to castles. To what extent were reporters and squatters complicit in the production of the housing crisis as spectacle on the very visible urban periphery? Beyond indications that the squatters used the physical landscape of their shantytown to shape its interpretation, a clue can be found in the descriptions of the people themselves. Besides a general openness among the residents, the "president" of the Kottbusser shantytown emerges as a consistent personality and the source of much of this performance of Berlin humor, as reported both in the *Gartenlaube* and by the *Hamburgische Correspondent* and reprinted in papers across the country. The *Bonner Zeitung* reported that Schmidt, a carpenter, had constructed the first hut "in the open field" by the Badbrücke. Taking his status as the most prominent member of the community, he held court and staged festivities. He invited members of the shanty community to celebrate the baptism of his son, born in Barackia and named in honor of his birthplace "Freifeld Schmidt." A carpenter by trade, he was teasingly referred to in the press both as "Schmidt the first" and "President Schmidt"—mingling monarchical and republican imagery, sometimes in the course of a single paragraph. The tongue-in-cheek account is full of references to possible treaties and trade between Barackia and foreign nations. In front of the "presidential hut," temporary tables were constructed from spare boards, which were then covered "in the absence of damask, with white sheets." The tables were laid for coffee "since the wine, according to the President, was desperately needed in the friendly neighboring state to host the Italian Crown Prince." The first toast was for the Kaiser. But in the midst of the frivolity, the settlement was also under threat. As he celebrated his son's birth, President Schmidt sought "to keep the threatened execution-by-rent from the Berlin magistrate at a respectful distance."[66] Press coverage of the shantytowns depicted humor as a form of resistance—a form of resistance perhaps distinctive to the acerbic wit for which the people of Berlin were known. This deployment of humor in the settlements, particularly as recounted in the rollicking stories of the Hamburg correspondent, evokes the darkly playful images of that most famous documenter of the Berlin working class, Heinrich Zille.

Resistance did not come only in the form of humor and the staging of the settlements for observation by community outsiders. The shanty dwellers also organized more direct forms of political protest. When police cleared the first of the shantytowns at the end of July 1872, rowdy crowds gathered to protest the action.[67] When Guido von Madai replaced Lothar von Wurmb as police president the next month, he announced the intention to clear the rest of the shantytowns by September 15 at the latest. Police had learned from the earlier protesting crowds and were probably made more cautious by the recent eviction riots as well. They planned to carry out the next evictions at night. But the shanty dwellers too had learned from their experiences and sought to protest evictions and advocate for their right to a home in new ways, sending delegations and petitions to several different branches of government. First, a delegation of men from the Landsberger Gate shantytown visited Police President Madai to plead for a deferral of their eviction. When this was unsuccessful, a delegation of women visited Mayor Hobrecht's office to try their luck with him. Finally, Albert Haack, a shoemaker, sent a petition to the king, listing his address as "2nd row, 1st shack." His petition became an oft-cited document in histories of Berlin housing, including in the later editions of Adolf Damaschke's book *Die Bodenreform* (Land reform).[68]

Police may have been concerned about the link between the shantytowns and working-class radicalism, but the press routinely depicted Berlin's peripheral shantytowns not only as demonstrative of the German national character but also as productive of patriotism. Many reports emphasized that the flags fluttering above the orderly shacks of the "new Berlin" were the red, black, and white, and the imperial eagle.[69] The shantytown dwellers were complicit in constructing this particular image of their settlement, not just in the flags they flew but also in the naming of their provisional streets: Empire Street, Citizen Street, and Colony Street. Socialists found this disappointing. Although the shantytowns might have provided an opportunity for new forms of working-class organization, instead the squatters seemed disappointingly apolitical, nationalist, and petit bourgeois in their aspirations.[70]

Social democratic desires for working-class organization and the middle-class desires for control converged in dislike for wild working-class settlements and spontaneous self-help movements.[71] That the squatters did not easily fit into an existing political movement does not mean that their actions were apolitical. Anarchist housing theorist Colin Ward argues that "there has always been a distinction between squatting as political demonstration . . . and squatting as a personal solution to a housing problem. In the first instance the intention is, for propagandist purposes, to be noticed. In the second, the hope is to be inconspicuous and to blend into the landscape." While the former rejects rent, the latter "actually hopes for the security of a rent book."[72] The Berlin shantytowns of 1872 demonstrate that such a clear distinction is untenable. The squatters in these settlements both performed their squatting for the public and also used it as a solution to their own housing crisis. Although most available sources are secondhand accounts, there is evidence in these accounts that the squatters' performance deliberately toyed with the confusion between opportunity and poverty inscribed in the space of their new homes.

ALLOTMENT GARDENING AND THE SELF-HELP MOVEMENT

Much of what appeared novel in the provisional urban landscapes of the shantytowns during the opening years of the German Empire would become a more familiar part of the urban landscape in later years, but under another label. In the opening shots of the movie *Berlin: Symphony of the Great City* (1927), the camera accompanies a train as it rushes into Berlin. Rapid shots of tracks, wheels, and telephone poles establish the rhythm and speed of a railway journey. Then, the camera opens up to the landscape along the route. Fields, trees, occasional houses, a bridge, a sign: "Berlin, 15 km." Suddenly a landscape of tiny huts and garden plots breaks through the trees, as cranes and larger buildings loom in the distance. They pass quickly, and in a flash the landscape turns industrial—a mine—and then Berlin's characteristic five-story apartment buildings appear in the midst of construction. Here, situated as a buffer zone at the frontier between the rural and the urban, is another landscape of huts, fences, and gardens. This is not a makeshift shantytown but a settlement of garden colonists (*Laubenkolonisten*)—the allotment gardeners who by the 1920s had become an established part of the German urban landscape and who played a particularly significant role on the periphery of Berlin. Like the 1872 shantytown settlers, garden colonists practiced a kind of settler urbanism.

The landscape of allotment gardening has frequently been interpreted as one of ambiguity. Descriptions of these urban peripheral garden plots play with the confusion between need and comfort, production and recreation, provisionality and permanence. In allotment garden settlements, one might encounter warming huts styled as tiny Swiss chalets or permanent homes made from salvaged half-ruined train cars dragged into place amid the fruit trees and vegetables.[73] As part of the urban peripheral landscape, allotment gardens have long been difficult to categorize by socioeconomic class, spatial organization, or use. The confusion is not just perception. Like the shantytowns sometimes claimed as an early exemplar of later "wild" garden plots, the garden settlements that became a familiar part of Berlin's twentieth-century landscape mimicked both middle-class suburban homemaking and the villa aspirations of the upper classes. They had their origins, however, not just in bourgeois reform movements but also in the spontaneous self-help efforts of the working classes to escape the pressures of the Berlin housing market and to supplement meager incomes with self-grown produce. In the late empire and the Weimar Republic, the Berlin city government struggled to use peripheral garden settlements as a safety valve for urban tensions in a crowded city, at the same time keeping a check on these potentially uncontrolled and unruly spaces of working-class autonomy. The Social Democratic Party also later encouraged workers' gardens for both health and recreation.[74]

The characteristic small urban garden plots, still present across Germany today, are most commonly known as *Schrebergärten*, but the modern allotment gardening movement had multiple German origins beyond the work of Daniel Gottlob Moritz Schreber in Leipzig (see chapter 1). Diverse origins betray different motivations and goals for the provision of small plots of garden land to city dwellers. In the early nineteenth century, garden land parcels served as compensation for the millions who lost their rights of land use in the process of enclosure. In the 1830s self-tilled vegetable plots spread across Germany as a form of urban poor relief. In the mid-century, allotment gardens gained a pedagogical justification in addition to being simply a form of economic support for the poor: garden work helped educate the urban poor in thrift and hard work. Schreber's own version of the urban gardening movement stemmed from the Turners and focused on exercise and the importance of nature and physical activity to the well-being of children. Later, urban gardening movements provided spaces of rehabilitation for tuberculosis patients and veterans.[75]

Distinct in the German allotment gardening movement, the Berlin branch originated as a form of spontaneous self-help. In the 1860s agricultural crisis and housing pressure led to the development of settlements of garden colonists. In 1862 a number of former residents of the city rented small allotments in the Schlächterwiesen outside Kottbusser Gate where they built small sheds and planted vegetables.[76] According to a chronicle of the Berlin allotment gardening movement, the settling families had migrated to Berlin from the countryside and had resorted to this semirural way of life when they had been unable to find a home in the city. Early gardens in Berlin's east and north were dominated by factory workers and small craftsmen from places like East Prussia or Saxony.[77]

Historian Hartwig Stein sees these early gardens as a move of resistance against the Hobrecht plan, though sources for this claim are scarce.[78] These early movements had no organizations and emerged as a true spontaneous self-help movement. The connection between the shanty settlements of the 1870s and the Berlin garden colonists (*Laubenkolonisten*) is difficult to establish. While later histories of Berlin allotment gardening claim the Gründerzeit shantytowns as part of their history, contemporary accounts do not mention this. Instead they describe gardening more as an afterthought than as a primary motivation for the settlements. Regardless of intent, it is clear that early "wild" garden colonists and the shantytowns of the same years stemmed from similar motivations and responded to the same urban pressures. Even if not part of a conscious movement of resistance, gardening was a significant way in which ordinary citizens were able to shape peripheral urban environments.[79]

After the 1870s the informal "wild" garden settlements were joined by more organized Berlin urban gardening efforts. From the 1870s through the 1890s Berlin experimented with potato patches to provide some form of urban poor relief. In his history of the German allotment gardening movement, Stein argues that—as envisioned by city officials such as A. Dreitzel, the Berlin city representative and head of the city administration of the poor (*Armendirektion*; after 1872, *Armenverwaltung*)—gardens for the poor were a form of rationalized poor relief combining social conditioning, control, and support. Dreitzel wrote of the advantages of giving the poor access to potato patches, which would both diminish their reliance on welfare and also improve their health. By the mid 1890s there were forty thousand allotment garden colonists in Berlin. Use of potato patches as part of a system of poor relief continued in Berlin up until 1897, targeting working-poor families with children rather than the most destitute of Berliners. At about the same time as the city was abandoning urban gardening as part of its organized poor relief, the Red Cross began a new partially privatized system of the same, more focused on health measures and excluding alcohol.[80]

Both shantytowns and garden colonies on Berlin's periphery in the 1860s and 1870s were made possible by the way the city subleased land earmarked for future development. Most Berlin allotment gardens could be found on city land, with only a few on church land or private land owned by land cooperatives. The city auctioned off the right to lease out the land to middlemen, who then subleased those plots for gardens and temporary structures as an intermediate use until a future date when the land was either to be developed by the city itself or to be privately developed in a more permanent way.[81] These leases were, in other words, short-term and uncertain. This meant that the middlemen to whom the city auctioned off leasing rights had clear incentives to run up profits on the land subleased to garden colonists while they could, not knowing when they would lose the land. They did so not just through rents but also by offering additional goods and services to the colonists, especially the sale of drinks through a site canteen to those tending their gardens on the weekends. This led to concerns about drunkenness and immorality in the semi-informal garden colonies, resulting from the manipulation of the colonists by the exploitative middlemen who could kick them out at any time.[82] This background makes even more significant the seizing of control over a landscape of recreation, urban excursions, and drinking by the nearby 1872 shanty dwellers.

The allotment gardening movement exhibited many of the same tensions between middle- and working-class control and reform priorities as did other branches of the worker self-help movement. Just as in the public reaction to the Berlin shantytowns, the allotment gardening movement as it grew experienced the same anxiety over the need to clearly distinguish between independence expressive of lower-middle-class striving and the "self-help" of working-class desperation. Damaschke rejected names like "workers' gardens" out of the fear that they would prevent the spread of the allotment garden movement in wider circles of the middle class.[83] This is why, to this day, the allotment gardening movement in Germany is more closely associated with its Leipzig rather than its Berlin roots. Leipzig's Schrebergärten (also known as *Familiengärten*, or family gardens), associated with middle-class reform movements aimed at improving working-class ethics and bodies, were much more appealing to a broad public, and cleared the allotment gardening movement of any class-specific language or association with an independent working-class self-help movement. Micheline Nilsen observes that these class tensions found expression in aesthetic differences as well. She documents the aesthetic criticism of wild working-class gardens found in Berlin by the organizers of the movement who preferred planned, orderly, middle-class garden settlements.[84]

Demands for space for allotment gardening grew into the next century, alongside more familiar demands for public green space. The years around the turn of the twentieth century saw the formation of local clubs and organizations for allotment gardeners, such as the Verband der Laubenkolonisten Berlins und Umgegend (Association of Garden Colonists in Berlin and Its Surroundings) in 1897, followed by national and international organizations to protect gardeners' interests. The movement was given new urgency and impetus by wartime gardening efforts. After the First World War, national legislation regulated small garden management and instituted rent controls, and the creation of a "Greater Berlin" with much more expansive boundaries brought new green spaces, including six thousand hectares of allotment garden settlements, within the city's borders. These large allotment garden settlements became partially formalized as a solution to ongoing housing shortages, but Weimar-era laws reflect the city's ambivalence toward the garden settlements as semi-permanent housing. Drawing the line between garden shed and primary residence proved difficult. On the one hand, the government subsidized the construction of sheds that met certain requirements of permanence, but other laws sought to limit permanent residence in the garden sheds. The semi-formalized allotment garden neighborhoods gained a reputation for disorder and drunkenness. By 1933 there were 120,000 permanently inhabited garden allotments. This meant that the majority of Berlin's allotment garden plots had become permanent residences, accounting for around 3 percent of the city's total population.[85]

The landscape of these new kinds of urban frontier settlements became the topic of popular cultural representations that examined the way suburban settlement living straddled divisions between the middle and lower classes, between work and leisure, between urban and rural, and between permanence and impermanence. In Hans Fallada's iconic novel of humiliated white-collar destitution in Weimar-era Berlin, *Kleiner Mann—was nun?* (Little man—what now? 1932), illicit long-term living in a garden plot hut provides the protagonist and his wife with a respite from the city, representing their spatial exclusion from its society and economy.[86] Fallada's novel explores the profound ambivalences of these spaces, which existed so far outside ordinary urban life that the moral guardrails of the protagonists' petit bourgeois existence seemed irrelevant. In this fictional account, the garden settlements are spaces of potential political thuggery and low-level property crime in the form of gleaning activities, such as the stealing of wood from the surrounding area.

Fallada's protagonists resort to garden hut living from force of circumstance, but the later Weimar years saw the emergence of the "urban edge settler" (*Stadtrandsiedler*) as a distinct type with its own identity, detailed in guidebooks and even magazines.[87] As Barry Jackisch has noted, this movement in some ways made an easy transition to National Socialism. Although the Nazi state rejected the movement's emphasis on grassroots activism and independence, its antiurbanism, love of land, and valorization of the small farmer were easily adapted to Nazi ideology, and allotment gardening organizations quickly coordinated within the Nazi state.[88]

Although (or perhaps because) unplanned settlements were comparatively rare in Berlin, in contemporary discussions of Berlin's growth their appearance took on outsized importance as both positive and negative indicators of the city's health. To include provisional and improvised dwellings in the narrative of urban expansion is to acknowledge the role of working-class residents as the cocreators of urban space, though this participatory role extended far beyond the bounds of a few short-lived shanty dwellings. Placing informal settlement within the context of other self-help movements, we can see the emergence of modern urban planning practices in the 1860s through the 1880s as a series of struggles over the rights of urban residents to make their own urban spaces.

Looking back on the history of the allotment gardening movement in 1931, the Reichsverband der Kleingartenvereine Deutschlands (Imperial Association of German Allotment Garden Associations), itself only ten years old, identified the housing crisis of the years around 1870 as the motivation behind the Berlin movement: "These catastrophic circumstances became the driving social force for the creation of proper allotment garden settlements [*Laubenkolonien*] that grew from the earth like mushrooms to take in the homeless, much as we observed after the last war." Understanding the link between Berlin's housing crises and informal settlements of the 1860s and 1870s and the twentieth-century explosion of garden settlements as a safety valve for later urban ills allows us to see some of the deeper continuities in the role played by informal settlement as a self-help solution to pressures on the urban poor.[89]

Gardening was one traditional use of urban peripheral land that was recovered in a new form in the context of the rapidly growing industrial city of the late nineteenth century. So-called wild garden settlements at the urban edge became a familiar part of a landscape that also included contested use of recreational land and spaces of nature (such as forests), as well as other kinds of temporary structures and low intensity forms of land use that also recalled an older model of the wasteland commons on the urban edge. Such practices—both age-old and appearing novel in the industrial context—formed another part of the modern city's "shantytown frontier": a space of provisional and shifting use where the urban poor and lower middle classes strove to use nature to make themselves at home in the city. Artists such as Zille and novelists such as Fallada documented Berlin's wild periphery from the end of the nineteenth century into the 1920s and 1930s when makeshift use of land that was neither clearly public nor clearly private but part of a new kind of urban commons recalled the themes of urban/rural confusion and class conflict already evident in the shantytowns of the Gründerzeit.

COMPETING VISIONS OF NATURE AROUND 1900 IN FOREST AND WASTELAND

At the same time as illustrated magazines trumpeted the appearance of a "New Berlin" of surprisingly orderly shanties in the east of the city, a plan was proposed much more quietly for another kind of "New Berlin" in the city's west. This settlement too was to offer a safety valve for metropolitan tensions in a suburban retreat to nature; this retreat, however, was not planned for the urban poor but rather for the city's wealthy. The proposal came from the so-called Napoleon of real estate speculators, Johann Anton Wilhelm von Carstenn. Writing to Chancellor Bismarck and Kaiser Wilhelm, Carstenn proposed a villa settlement encompassing the cherished Grunewald Forest between Charlottenburg and Potsdam on the Havel River to Berlin's west. While working-class Berliners constructed makeshift homes on the other side of the city, Carstenn suggested that the best solution to the current tensions caused by the housing crisis was for the city's elites to leave the city center behind to the poorer classes, relieving housing pressure and allowing rents to fall. He feared a future Berlin crowded by working-class suburbs that might erupt into revolution at any moment.[90] For all the differences between Carstenn's proposed "New Berlin" and the shadow shantytowns constructed on the other side of the city, these two ideas for the future of the city shared a number of significant common features. Both visions figured a retreat to nature on the urban edge as a solution to contemporary urban ills, and both envisioned a return to lower-density settlement. If the shantytowns made poverty too visible and urban segregation by class too extreme for many observers, Carstenn's villas formed the other side of that version of the city—isolating and displaying urban wealth on the periphery—but they came as an expression of the very planning and speculation that contemporary observers saw the shantytown Berlin as rebelling against.

The Grunewald settlement was never built, but real estate speculators such as Carstenn and Heinrich Quistorp successfully developed a number of villa settlements on the edge of Berlin in the boom years of the Gründerzeit. Carstenn founded such settlements in the southwest fringes of Lichterfelde, Friedenau, Halensee, and Wilmersdorf. After 1871 nearby villages experienced explosive growth, fueled by the shortage of space in the capital and the construction of train lines. Charlottenburg and Schöneberg became large cities in their own right. Only villages far from the city succeeded in retaining their previous structures, creating a new nostalgia for the urban hinterland of forest, field, and limited settlement. The development plans of the dozens of joint stock companies and mortgage banks created in the previous decade dissolved in the crash of 1873 at the end of the boom—again mirroring the explosive growth and disappearance of the shantytowns.[91]

Placing Carstenn's failed plan for the Grunewald Forest on Berlin's edge alongside the shantytowns of east Berlin makes clear how these different proposed versions of urban expansion engaged not just questions of planning and class but the relationship between nature and class in the city. In an era of seemingly unbounded urban growth, the urban edge remained a space of retreat to nature. Jeffrey K. Wilson recounts Carstenn's plans from his correspondence with Bismarck and Wilhelm as a piece of the history of the Grunewald Forest as a site of emerging environmental activism in imperial Berlin. Carstenn sought to take the onetime royal hunting preserve and turn it to use as an elegant retreat for Berlin elites, and Wilson details how left-liberal politicians, a range of public associations, and the popular press worked to save the peripheral forest for the people of Berlin. They believed that the forest as a natural resource belonged to the people of the city, and that it could and should be mobilized against a range of modern urban ills. Wilson's account traces nineteenth-century ideas about green space and whom it should serve, evolving from a space for the king and nobility to a space for the German people as a whole. This idea did not have a clear political affiliation but, rather, united conservative and liberal strands of German social thought.[92]

The conflict over what to do with the Grunewald reflected a clash between different planned visions of the urban edge—as a green space for recreation and access to nature and as open land for planned housing developments.

Urban reformers of the same years, such as Countess Adelheid von Dohna-Poninski, looked to the possibility of a "green ring" around Berlin, modeled on cities such as Leipzig that had made better use of their former fortifications land. Dohna-Poninski envisioned urban green space as a force for social and economic integration, where Berliners of all ages and classes could make use of the land. Her particular concern was the excessive growth of the city's population among the lower classes, whose "unplanned" (*planlos*) housing and neighborhoods posed a threat to the morality and safety of the whole city. She believed that planned, ordered access to open space for recreation and land for gardening would help restore community, family, and health of Berlin's workers.[93]

In the standard narrative, the urban edge of Berlin entering the twentieth century was increasingly characterized by stark contrasts between open land and new dense settlement. As British planner Patrick Abercrombie put it in 1914, while the typical city grew in a landscape of "small roads and peddling suburban houses," Berlin instead expanded by pushing "wide town streets and colossal tenement blocks over the open country, turning it at one stroke into a full-blown city."[94] The Grunewald people's park (*Volkspark*) was part of a new green hinterland for Berlin, but among the many uses anticipated for its open spaces were not included dwelling, gardening, or gleaning. This salubrious space of nature was instead to be reserved for health and recreation, not sustenance. The "primeval forest" on the city's edge provided a nearby wilderness for urban outings—part of a new recreational hinterland of the city providing parks, forests, beaches, and other destinations for weekend excursions for city dwellers cooped up in the crowded city during the week.[95] Instead of a space for villas, the Grunewald forest and other urban peripheral green spaces became a space for a very particular vision of the consumption of nature for leisure and health.

Contrary to the rhetoric of early environmental activists who asserted a popular right to nature, the Grunewald was no primeval forest but, rather, the product of economic forestry methods born in the eighteenth century. The idea of the Grunewald as untouched nature expressed class-specific ideas about what constituted appropriate use of the open spaces on the urban edge. Categorizing the Grunewald as an unchanging landscape preserved from Germany's deep forested past meant protecting it from economic or practical use: the forest was there to be enjoyed, a source of recreation and physical and psychic reinvigoration through contact with nature. In a space reserved for leisure, contemplation of nature, and the business of health, other uses of the forest that might suggest the precarity of urban life were controlled and contested. Gleaning, settlement, and small-time commerce (especially the selling of drinks and pretzels) were prohibited in a highly class-specific vision of nature on the periphery.[96]

One startling set of images of the working-class periphery, and particularly working-class uses of the Grunewald Forest, comes to us from Heinrich Zille, suggesting a link between the landscapes of nature and growth on the periphery and creating an urban periphery that remained ambiguous and isolated from clear narratives of either urban progress or pastoral preservation. Zille's visual representations of the city were deeply rooted in his personal experiences as a migrant to Berlin in the years of the Gründerzeit. He grew up in the working-class east as a Saxon immigrant and later, as a successful artist, lived with his wife and children in a respectable middle-class neighborhood in the west of the city. He is better remembered for his drawings of the teeming inner courtyards and taverns of the city's workers' quarters, but Zille was also fascinated with the empty spaces at the city's edge. He chose to live on Berlin's outskirts where he had a view of open space beyond the built landscape.[97] In his photography, he showed particular interest in the ragged edge between planned and unplanned spaces. He photographed multistory apartment complexes jutting out into the empty fields and beyond them a landscape of transition and precarity, where roads end abruptly amid wild trash heaps, packs of children roam, and wasteland gleaners collect sticks to keep their fires burning, barely beyond the shadows of those newly constructed Mietskaserne. Zille's photography celebrated this ragged urban periphery as a kind of rediscovered wasteland commons, resurrected from the lost common land of an earlier era.

In 1897 and 1898 Zille took four series of photographs of women and children collecting sticks in the area around the Grunewald Forest and the West End colonies of the rapidly growing city. This was near his new home and the photography studio where he worked. All of these images include either a single woman or a group of several women carrying loads of sticks back from an unseen source. In a few, children accompany the women. In some images they push carts alone or in pairs, in others the sticks are loaded into a baby carriage. Sometimes women carry their scavenged wood in baskets or bundles, sometimes on their backs. In most of the photographs the women are seen moving toward a line of apartment buildings, with the characteristic form of the Berlin Mietskaserne in the distance. In most of the photos, the women's backs are to the camera. In some they are moving, in others at rest, apparently taking a break in their journey (see figure 4.2).

These photographs resemble paintings such as Max Liebermann's *Holländische Reisigsammler und Holzfäller im Wald* (Dutch stick collectors and woodcutters in the forest, 1911) or Franz de Beul's *Die kleinen Reisigsammler* (The

4.2 In several of Heinrich Zille's photographs of women collecting wood on the western edge of Berlin, the women are accompanied by children, and the sticks conveyed in baby carriages. Heinrich Zille, *Frau und Kind mit reisigbeladenem Kinderwagen, im Hintergrund die Knobelsdorffbrücke, Berlin-Charlottenburg*, 1897 (reprinted 1978), silver gelatin photograph, 20.2x26.4cm. Copyright Stiftung Stadtmuseum Berlin.

little stick collector, ca. 1900), but paintings in this familiar genre are typically placed, as one would expect, in forested landscapes.[98] Notably, in Zille's images the sources for all these sticks are largely absent. Unlike the paintings, few trees appear. Sometimes a line of trees is visible along a road in the distance, but these are planned trees, a part of the city—not a likely source for the large heaps of sticks in the women's packs and carriages. The source of the fallen branches is left out of sight—and the activity, for the barrenness of the surrounding landscape, is left somewhat mysterious and incongruous. The women are presumably coming from the nearby Grunewald Forest, though within the image it is not clear where they gathered the sticks in their cart.

These photographs contain an implicit tension, because the women's activity seems to be not entirely of the space they are in. They are in a sense images of a border: visual representations of the edge of the city as the limit between ordered and unordered space and also as a boundary between intensive and extensive uses of space. You can see the linear, ordered space of planted decorative urban trees, fences, and new housing developments in the distance. The foreground space is both empty and less clearly segmented. The sandy ground visible in many of the images evokes familiar characterizations of Berlin's hinterland as infertile, desert- or steppe-like. The women in the foreground are outside the ordered space of the city but oriented toward it. In most of the images the figures are traveling not on a road but simply through open land on the city's periphery. Choosing to photograph the women facing away from the camera emphasizes their physical toil, presenting them less as individuals and more as part of a landscape, highlighting the likely repetitive frequency of such treks back and forth.

These are also almost certainly technically images of theft. Stick collecting was a natural activity in the older spatial arrangement of the city, but now it had become a transgression on private property in the new arrangement of the city. Much as the images of Berlin's shantytowns and wild gardening settlements, Zille's photographs also depict the ways in which urban life had become difficult to the point of impossibility and illegality for Berlin's working classes. Referencing the activity of stick collecting, these images claimed the urban edge as a wasteland commons—a space for collective, extensive (that is, low-density) uses such as collecting firewood. Part of the historical economy of the commons, such gleaning and foraging had long provided women and children ways of contributing to their households outside official or formal exchanges.[99] Zille, too, presents stick collecting as an economic activity performed by women on the margins of the urban economy—but oriented toward it.

Zille's images of a wasteland commons on the urban periphery challenge clean divisions between public and private space and emphasize a working-class culture of improvisation and accommodation, while providing concrete portrayals of a space between the urban and rural. Zille's photographs and artwork betray a fascination with improvisation, impermanence, and common land on the urban edge. The urban peripheral landscape of Zille's images is part of the same world of precarity and mixed use as shantytowns and allotment gardens. In an urban hinterland increasingly dominated by dense settlement and protected cultivated spaces of nature for recreation and retreat, shantytowns, wild garden settlements, and Zille's stick collectors all represent spaces of precarity and provisionality on the urban edge. Where, after all, did the scrap wood and the building supplies for a landscape of improvised shanties and garden huts come from but from the marginal economy of gleaning and making-do revealed in Zille's images? This landscape shared with Carstenn's New Berlin an idea of return to nature but challenged the rigorous planning and privatization of that nature.

Including wild settlement in the account of the historical development of the German city provides a corrective to the center-out and top-down approaches to city development history that have dominated the picture. Unplanned settlement can incorporate two kinds of use of space that are not always clearly distinguished from each other: the transformative uses of existing spaces in unanticipated ways and the construction or appropriation of new spaces. Incorporating these so-called wild forms of settlement in our history of the city gives us a richer understanding of the origins of the urban environment and the diversity of actors involved in the active creation of urban spaces. This approach to the city has profound advantages for studying urban growth in a global and comparative context. Looking at the German city in terms of a kind of settler urbanism that linked housing movements, shantytown settlements, allotment gardening, and the later Stadtrandsiedler movement provides a common structure for thinking about the landscape of the expanding German city of the nineteenth century together with a global landscape of German settler diaspora. In the United States, especially in New York, urban shantytowns were also spaces of migration and are often associated with German immigration. One function of press coverage associating the Berlin periphery to the American West was to connect working-class responses to urban housing crises in the center of the new empire with the collective myth and imagined experience of migration.[100]

Berlin's urban peripheral shantytowns provide a concrete representation of the advancing urban frontier: a vision of city growth as progress and invention. As the American frontier myth turned change into a core and semipermanent feature of the American character, urban observers in the same era sought a fixed image of the city's shifting frontier—a visual identity rooted in change. The shantytowns of 1872, placed "just outside" now absent gates, again rendered the urban horizon, the horizon of the city's expectations, visible. This is what made the shanties' use of space radical, alternately disturbing and full of potential for the city's future. They represented the emergence of the lower middle class in the uncomfortable, half-natural, half-industrial spaces at the edge of the city. Like Zille's stick collectors and lonely apartment buildings, the shantytown frontier made the indeterminate landscapes of the urban periphery visible. Like the tax wall in its moment of deconstruction, it made the process of urban growth into a spectacle: a public event, performed by residents and by the press for consumption not just by the Berlin public but also by a national German audience, and even observers abroad, offering an alternative vision of German home and place.

5

Urban Histories and National Futures in the German Empire

In 1907 the German sociologist Werner Sombart observed that his contemporaries' understanding of the city had not kept up with the rapidly changing urban environment around them. Asked what the word "city" meant, he claimed that most Germans would call to mind not a definition but a picture: "the picture of a settlement of many people in houses and streets, if possible surrounded by walls and battlements, a settlement that is sharply set off from the countryside." This was not a picture of just any German city, he observed, but a very specific one, "perhaps Nuremberg." And it was not just any Nuremberg either, but the Renaissance city "as Albrecht Dürer had drawn it."[1] In search of a functional definition of "the city" to guide contemporary social policy, Sombart found instead a Renaissance illustration. Although nineteenth-century magazine writers and artists might have sought to provide concrete images of a changing urban frontier in the present, Sombart feared that the imagined walls of Germany's imagined cities might hinder the real economic growth and political maturation of the German nation.

By the time Sombart penned these words, much had changed since the Gründerzeit era of wild growth. In 1871 less than one in twenty inhabitants of the German Empire lived in cities larger than one hundred thousand residents. By 1910 the ratio reached one in five. New infrastructure and planning initiatives transformed the physical landscape of Germany's cities. But just as the Renaissance walled city seemed to be disappearing for good, the historical legacy of Germany's many cities and towns emerged as an important issue in social scientific and popular discussions of the German Empire's political and economic developments. The nation was gripped by a fascination with history, from historical fiction to historicist architecture, and also by a special interest in how changing social and economic structures were transcribed into spatial relationships.[2] Concern for historical preservation and a nationwide interest in the appearance of the German cityscape drew attention to the durability of urban institutions. The elevation of cherished monuments of urban architecture as the visual vernacular of a united German culture forged a bond between the new nation and the urban communities that were the natural habitat of German liberalism.[3]

In the minds of many liberal politicians, reformers, and intellectuals in the German Empire, cities fostered uniquely urban values of cultivated civic-mindedness, independence, and enterprise that were essential to the nation's political and cultural health. As the work of Jennifer Jenkins on Hamburg has shown, the ideal of cultivated urban public culture in the late nineteenth century linked new conceptions of liberalism and citizenship to older, more localized definitions of urban identity. Urban fairs like those held in Berlin and Dresden in 1896 and 1903 not only celebrated the contemporary big city but also recalled the riches—literal and cultural—of Germany's medieval and Renaissance urban centers. The Dresden exhibit sought to root the achievements of the modern city in a celebrated urban past—as evidenced by the apparent incongruity of an exhibit catalogue cover with a picture of a walled city guarded by a medieval knight.[4] It was clear that the historical city continued to hold a central place in both the visual identity and civic culture of the German Empire.

This celebration of the urban past in the empire was a conflicted one, however. Elsewhere in his work Sombart mobilized a different picture of the historical walled city. In *Die deutsche Volkswirtschaft im neunzehnten Jahrhundert* (German national economy in the nineteenth century), published in 1903, he described the radical transformation of the German landscape over the past century from a place where travel was difficult and the terrain wild. He encapsulated this transformation by describing how the borders of a city such as Berlin had changed. A century earlier, he wrote, the city was isolated, as if surrounded by a desert. But instead of welcoming the thirsty and dust-blinded traveler, its walls were cloaked in the stench of trash, excrement, and the rotting carcasses of domestic animals thrown out of the city. Miserable extramural dwellings ringed the walls. Excise officials beset all who approached with endless questioning about where they had come from and where they were going. This image was the shadow side of Nuremberg's Renaissance charms and represented the danger of an urban identity that remained mired in the past: a national space in which freedoms and services were jealously protected, where public mindedness ended at the city's edge, and a voyage from Berlin to Dresden, by Sombart's cartoonish account, equaled a trek to Siberia in risk and difficulty.[5]

The power of Sombart's negative image of a walled Berlin rested on its ability to evoke a story of dramatic transformation. The modern city with its open frontiers, train stations, and clean streets was the antithesis of the perceived stagnation of the eighteenth-century city, closed and decayed. Sombart's description evoked an enduring fear of the stasis thought to have gripped the German city between its Renaissance heyday, represented by Dürer's Nuremberg, and its mid-nineteenth century rebirth as a growing industrial city—precisely the long arc of the German city's gradual overcoming of visible walls and barriers. How the idealized Renaissance city, characterized above all by the guild and the gate, had given way to the urban industrial capitalist economy—across the divide of an eighteenth century that had marked a transition for both the German city and the German national consciousness—was a crucial question for determining the character of both modern urbanism and the modern nation.[6]

These two historical images of the German city—the turreted Nuremberg of the sixteenth century and the garrisoned Berlin of the eighteenth—together represent the difficulty posed by the historical relationship between urban and national spaces in the German Empire. Insofar as Germans hoped that the nation would inherit the spirit of enlightened reform embodied in the idealized late medieval and Renaissance city, the premodern walled city evoked a shared and celebrated past for the educated middle classes. On the other hand, the city wall also represented the limitations of that project, suggesting that the principles of participatory self-government animating the historical city had succeeded only within the narrowly confined space of its walls. This tension shaped public debates in cities such as Cologne, Bonn, Düsseldorf, Munich, and Nuremberg over whether to remove city walls and gates or to preserve them as historical monuments. Depending on one's perspective, city walls and gates that remained standing into the final decades of the nineteenth century were either monuments to an ennobling heritage or oppressive reminders of the cramped conditions of past urban stasis.

These were the sorts of debates through which the organized preservationist movement emerged at the end of the nineteenth century. Rudy Koshar identifies historical building preservation as a new kind of "memory work" (*Erinnerungsarbeit*) for the German nation after 1890.[7] What is missing in studies of historical preservation as a late century middle-class phenomenon is the degree to which preservation was just one among many ways of viewing the built environment through its connections to the past. It is not how preservationists read the built landscape as evidence of the past that was new but, rather, their claim of expert authority over these debates and the assertion that the nation was a privileged context for historical memory. Historical consciousness played a key role in city making, far beyond the preservation of historical buildings, and had done so since at least the middle of the nineteenth century. Even when cities removed their walls and fortifications for new construction, these projects also engaged an understanding of the past and helped assert historical continuities. "The performative recovery of place and time" was, as Umbach argues, a distinctively modern project, and Germans "performed" the recovery of historical narratives by building as well as by preserving.[8]

Even when it came to historical preservation, a building might be seen as a witness to the past in different ways, corresponding roughly to Umbach's distinction between history and memory. If a building were taken as a document of a specific historical moment, one might construct a narrative accounting for the historical process of change connecting that moment to the present. If, on the other hand, a historical monument became emptied of its specificity, to represent instead a vaguely conceived premodern "before," it could not be connected to the present by a narrative of historical change and so was available only as a kind of mythical antithesis to the present. This, indeed, was what Sombart so feared in that idealized image of old Nuremburg. Freed from its context, the image functioned as a decontextualized and universally available depiction of the city: not as history, but as myth. How Nuremberg gained that status can be seen in the debate over the preservation of its walls in the first years of the empire.

WALLS AS HISTORY AND MEMORY IN THE BATTLE OVER PRESERVATION IN NUREMBERG

Among those German cities where planned defortifications garnered widespread notice, the most visible and talked about was Nuremberg. For many Germans the city of Hans Sachs represented what was best and most noble in German burgher culture. As a city of Protestant Franconia incorporated in Catholic Bavaria, as a jewel of German Renaissance architecture, and as a city with a rapidly growing urban underclass suffering from a near critical housing shortage in the 1870s, Nuremberg characterizes many of the most important themes that came into play in the preservation of city walls in the German Empire. Because of its iconic status, the debate over its walls also engaged concerns about the character of German national culture and the relevance of the urban past to the future of the young empire. Berlin may have been Germany's political capital, but it was Nuremberg that became the capital of an imagined German cultural nation (*Kulturnation*) in the nineteenth century.[9] Nuremberg offered an alternative to the Prussian capital, its turrets and narrow winding streets giving palpable shape to a large German nostalgia for the Holy Roman Empire.

Although Nuremberg's reputation for historical beauty went far beyond its ramparts to encompass the city as a whole, the walls became a convenient metonym for the city's value as an object of preservation. By turns both imposing and charmingly picturesque, Nuremberg's well-preserved fortifications inspired passion among artists, amateur historians, tourists, and preservationists alike. The conflict over Nuremberg's walls was particularly fraught because preservation concerned the nearly complete circuit of the city's old fortifications. The question of demolition versus preservation concerned not an isolated monument but the border itself. The struggle over Nuremberg's walls brought into sharp focus the perception of a conflict between urban growth and development and historical preservation.

The urban fortifications that survived into the mid-nineteenth century had been built in the fifteenth century and strengthened in the sixteenth and seventeenth (see figure 5.1). The city had dry moats, ramparts, and a wall nearly twenty-five feet in height that followed the contours of the hilly ground on which the city sat, under the shadow of its castle.[10] Five main round-towered gates provided the city's main entry points, also breached by the Pegnitz River, which divides the city. The great number of towers scattered along the wall's twisting course gave the city its characteristic profile. The city's cluster of turrets inspired naturalizing metaphors: in the words of many observers, they grew upward as if they had sprung from the earth.[11] Written accounts of the physical city were often mediated by reference to familiar pictorial representations: an old map or a woodcut.[12] Nuremberg looked both like an organic product of its landscape and also like a picture of itself.

Those early modern fortifications signified Nuremberg's onetime power and importance as an imperial city. Designed in part by Albrecht Dürer, they evoked the heights of German scientific and architectural achievement. Their

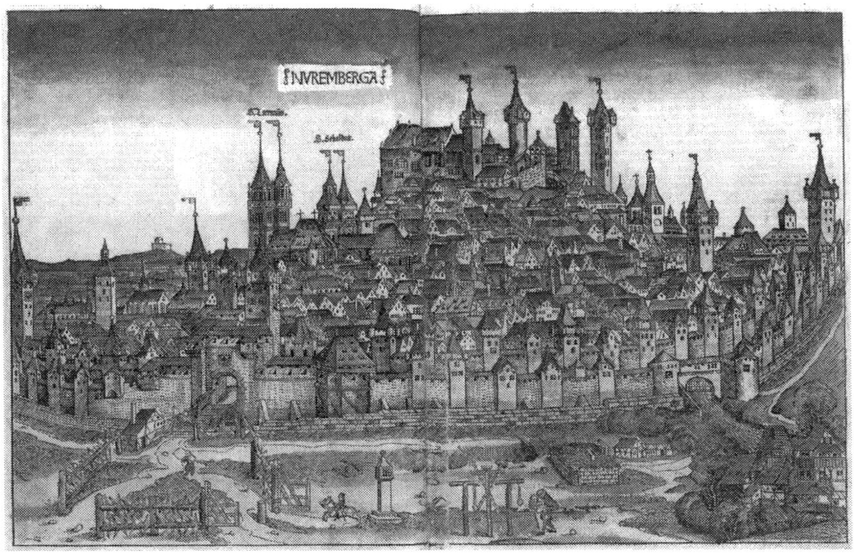

5.1 This is the earliest published view of the city of Nuremberg, from 1493. Michael Wolgemut and Wilhelm Pleydenwurff, "View of Nuremberg." Woodcut in Hartmann Schedel, Liber Chronicarum (Nuremberg: Anton Koberger, 1493), Museen der Stadt Nürnberg.

continued presence helped make possible the elaborate "historical drag" of the city's many folk festivals, which became a crucial part of the Nuremberg tourist industry.[13] Even outside of festivals in the everyday life of the city, the appearance of Nuremberg's well-preserved streets enticed many writers to imagine its inhabitants in old-fashioned dress. Woodcuts and watercolors showing picturesque views of the ivy-covered and charmingly dilapidated towers, walls, and ramparts were reproduced and sold in one of the many late nineteenth-century cottage industries produced by the popularization of romantic views of the past. The walls were also a popular subject for artists, such as Andreas Mattenheimer, Johann Heinrich Audenrieth, Max Bach, and Lorenz Ritter.[14]

City administrators and urban residents, however, saw the walls as increasingly burdensome. Traffic between the urban center and the expanding industrial suburbs increased in the 1850s and 1860s, and a housing shortage placed space at a premium. The Nuremberg city council and the energetic reformist Mayor Otto von Stromer began to push for the removal of large portions of their city's fortifications in 1868, just two years after the city lost its official status as a Bavarian fortification. Under the direction of August Essenwein, the walls' preservation became a cause célèbre of the Germanisches

Nationalmuseum (Germanic National Museum), which was founded in 1852 and located just outside Nuremberg's city center. The museum played a key role in securing Nuremberg's symbolic importance to the nation. Artists, architectural historians, and a community of preservation enthusiasts associated with the museum worked to bring this threat to national attention in both specialist journals and popular periodicals.[15]

Just at the birth of the Kaiserreich, Nuremberg's walls were for a number of reasons poised to play a particularly prominent role in a national historical consciousness also plagued by anxieties about unification. Accounts of Nuremberg as Germany's imagined cultural capital occlude practical conflicts like those over the city's battlements, and previous accounts of the battle over Nuremberg's walls have focused on local politics, leaving unclear why the issue had such purchase beyond Nuremberg in the national popular press and in the preservation and art communities.[16] What is needed is an exploration of both—to examine how local and national attention to Nuremberg's walls interacted in a negotiation over the relationship between history and place.

National interest in the fate of Nuremberg's built landscape rested on the city's special status as a German cultural capital. The magazine *Ueber Land und Meer* (By land and sea) declared that Nuremberg's built environment held special significance as it was one of very few places to which the whole country had a cultural claim. A Berlin-based architecture journal asserted that the walls testified to Nuremberg's historical importance while "recall[ing] at every turn the history of our entire German fatherland." Their characteristic silhouette simultaneously evoked a shared national heritage and also set Nuremberg apart as unique among German cities.[17]

For advocates of historical preservation, Nuremberg's iconic status gave the city a burden of responsibility toward the nation. Insofar as Nuremberg was a national treasure, the German people as a whole exercised a kind of proprietorship over the city, which could be used to challenge both the autonomy of its municipal government and the responsibility of the Bavarian state for its historical riches. The poet Hermann Allmers argued that the Bavarian king had vacated his right of ownership by neglecting his duty to protect this monument.[18] Within the city, national pressure to preserve the walls seemed to some to marginalize Nuremberg's contemporary importance as a city with a vibrant economy, industrial sector, and changing political landscape. The Bavarian state government employed similar logic in its own attempts to temper the enthusiasm of Nuremberg's city administration for demolition of the walls. The Bavarian general conservator in Munich, responsible for historical preservation throughout Bavaria, advocated the preservation of Nuremberg's walls by citing the city's responsibility to the "educated class in all of Germany and abroad." For advocates of a "large Germany" that included Austria, the close association of Nuremberg's walls with the legacy of the first German Empire served not to glorify the newest German Empire but, rather, to represent a larger German cultural tradition supposedly sacrificed in the drive to unification.[19]

Whether for the sake of the Kaiserreich or of a supranational Germanic culture, preservationists pitted the supposedly selfish, materialist, and short-sighted interests of the municipality against a larger cultural claim. In *Die Gartenlaube*, the illustrator Herbert König lamented the destruction of walls, fortresses, and gates in the face of the reigning "utilitarian principle" and beseeched local governments to recognize their responsibility for protecting these structures from the contemporary style, which valued growth and progress above all else. The so-called leveling times (*nivellierende Zeit*) demanded the sacrifice of the imperial city's beauty. The director of the Germanic Museum, August Essenwein, lamented that the walls were "old rags" when viewed with the "academic-bureaucratic wisdom" of philistines who sought only to make everything modern.[20]

As with the removal of Berlin's wall in the 1860s, articles about Nuremberg made reference to the tearing down of city walls as a kind of bloodless, slow motion revolution. But instead of looking back to 1848, more recent events in Paris loomed large in discussions of Nuremberg in the 1870s. In an 1872 article entitled "Eine unblütige Commune" (An unbloody commune), art historian Moritz Thausing framed the attempt to take down Nuremberg's walls as a declaration of class warfare. Thausing posited that the gratuitous removal of city walls was the characteristic form taken by German revolution—a slow-burning rage of a people against their own past, quieter but no less destructive than the more spectacular French style of revolution. The compulsion to destroy the physical landscape resulted from a chronic illness of a grasping, petite bourgeois philistinism, characteristic of the German urban middle classes and falsely clothed in populist language.[21]

Preservationists deployed the divisive rhetoric of revolution and class warfare, but they sought to associate their opponents with that divisiveness while themselves taking the moral high ground of selfless concern for the nation's cultural legacy. As he accused Nuremberg's progressives of strategically deploying a vocabulary of class war, Thausing appealed to that inflammatory language of class conflict himself. "The poor make the riots," Thausing wrote, "and the rich destroy the historical architecture."[22]

Those who objected to the removal of Nuremberg's walls also deployed a religiously evocative vocabulary to frame their case. Preservationists deemed the walls' demolition a "sacrilege." An 1873 article in the *Gartenlaube*, for example, described the destruction of Nuremberg's walls as "sacrilege against the reliquary chest of the German Reich." Essenwein compared the walls to a cathedral, in order to argue that any compromise agreement that preserved only portions of "the ramparts from which Hans Sachs sang the city's praises" would be "a death sentence against the whole that destroys its integrity." The walls formed a coherent structure just like the more obviously unitary structure of a cathedral. In the same article, Essenwein described how artists from across Germany made "pilgrimages" to Nuremberg to see its walls.[23]

In an empire fractured by religious tensions between Catholics and Protestants, religious language provided a framework for discussing questions of preservation and urban reform that drew on existing discourses and added a heightened sense of urgency. To see a clear battle pitting Catholics against Protestants in these debates, though, would be to miss the point. Evident instead is a fusion of Protestant and Catholic imagery. The walls are like a cathedral, but one from which the Protestant hero Hans Sachs sang the city's praises. The city within its walls was a reliquary chest and a place to which pilgrimages were made, but Nuremberg's significance was the secular celebration of German *bürgerlich* culture. In this way, Nuremberg's walls operated as a secular alternative to religious sites of pilgrimage, a site that represented a unified culture above confessional divisions.[24] Both Essenwein at the Germanic Museum and the editors of the *Gartenlaube* were engaged, albeit in different ways, in forging a unified national culture that brought together appreciation for Germany's historical heritage with a celebration of its present flourishing. They offered up Nuremberg's walls as a suitable symbol of that unified culture, while mobilizing religious imagery that evoked the most vital threats to its cohesion. They effectively raised the stakes of preservation by equating an attack against the walls with an attack against that very unity.

This rhetorical strategy is clear in an article on the "vandalism" of Nuremberg's walls that appeared in 1879 in the nationally distributed *Kunst-Chronik* (Art chronicle), which deployed religious imagery in a more explicit way. The article, authored by the Munich-based painter Jso Krsnjavi, compared the alleged fanaticism of the progressive case for the demolition of Nuremberg's walls with the religious enthusiasm of politically suspect Catholics. Krsnjavi accused Nuremberg's city government of revering the Nuremberg industrialist Lothar von Faber and his widely known city expansion plan as if it were God's work. The city councilmen, he mocked, seemed within Nuremberg to be "more infallible than the Pope." This formulation required a bit of mental gymnastics on the part of its readers, since the city council was composed of Nuremberg's liberal, Protestant ruling class.[25]

Outside of Nuremberg, resistance to the removal of the wall reflected anxieties about the conflict between particularist interests and a unified German culture. Locally, the stakes of the debate were different. Within the city, preservation appeared as a special interest, while the walls' removal offered a project in the interest of the general good. National resistance to Nuremberg's defortification peaked in the mid-1870s, preceding the local movement for preservation in Nuremberg, which did not gather steam until after 1877. In many ways, this local movement was itself the product of the national movement, rather than being a "home-grown" expression of local pride. The Gesamtverein der deutschen Geschichts- und Altertumsvereine (General Society of German Historical and Antiquities Associations) held a meeting in Nuremberg to celebrate the twenty-fifth anniversary of the Germanic Museum in 1877, the same year that the Nuremberg city government published its first comprehensive plan for defortication and expansion of the city. That year, the General Society resolved on the formation of a local Verein für Geschichte der Stadt Nürnbergs (Association for the History of the City of Nuremberg).[26]

Treatment of the walls' possible removal in the local association's journal, however, echoed little of the alarmist language that was found in the national press. In an initial article published by the association on this subject, local artist Max Bach sounded a note of melancholy over the disappearance of the walls but avoided an overt plea for their preservation. "The walls of Nuremberg, once the pride of the city," he wrote, "must more and more give way to the claims and needs of the present." Even if some stretches of the walls were preserved, he reasoned, who knew whether they would remain intact in the future? The walls' historical value was less reason to advocate for their physical preservation than for the urgency of recording their history and documenting their present-day appearance. That was the task that Bach set himself, both in historical essays and in a collection of sketches of picturesque views of the city and its ramparts (see figure 5.2).[27] For Bach, the "virtual" preservation of cumbersome historical monuments offered a way to reconcile respect for and celebration of the past with the fostering of the city's future development and growth. Seen from the perspective of the city, the walls were part of a history of development that could not be preserved as a picture but were, rather, part of a living environment. The walls were infrastructure. Although sometimes beautiful, encumbering infrastructure is not to be preserved but updated.

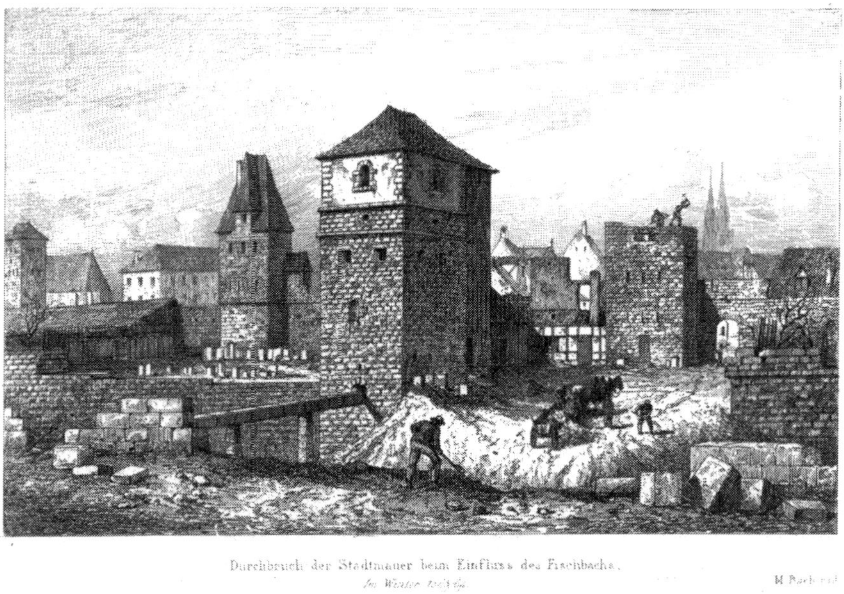

5.2 "Opening of the City Wall at the Inflow of the Fischbach." In Bach, *Architektur-Skizzen aus Nürnberg*, Columbia University Avery Library.

To the extent that there was an enthusiastic movement within the city itself for the preservation of Nuremberg's walls, one of its primary constituencies was the group that had perhaps the clearest reasons for fond memories of Nuremberg's status as a former imperial city: those descended from the patrician families who had governed the city until 1806. The descendants of these families made up the core membership of the Association for the History of the City of Nuremberg and published works on the history and appearance of the wall.[28]

The other local group that later emerged as advocates of preservation was the city's art community—some of whom took a less pragmatic view than Max Bach. In 1886 a group of fifty-two local artists and craftsmen, led by the Nuremberg painter Lorenz Ritter, submitted a petition to the city magistracy to plead for the preservation of the remaining portions of the fortifications. They appealed to the walls' aesthetic and historical value, but they also referred to the special interest of artists in preserving the structures many of them depicted in their work. They feared that their own professions, like the walls themselves, might "fall victim to material interests" as the "spiritual" and cultural identity of the German nation gave way to the economic preoccupations of a materialist society. Artists saw the walls as a symbol of the past rather than a surviving monument to a particular period in history. Failing to mention specific architectural features of the wall, they lament the potential loss of its romantic "dark green ivy coat."[29]

The idealization of the walls as an architectural landmark perpetuated an image of the walls that had little to do with their actual appearance. The moats and walls were not simply empty and ivy-bedecked but were also intensively used spaces. Photographs like those taken by Georg Schmidt in the 1860s and by his son Ferdinand Schmidt from the 1870s through the 1900s vividly depict how the city lived its daily life in, on, and around the walls and moats. Beyond picturesque views, they offered space for craftsmen to work, families to plant vegetables and hang laundry, and the city to store lumber.

From the perspective of the Nuremberg city government, claims of either a "small German" or a "large German" cultural proprietorship over their walls were perceived equally as unwelcome outside interventions in the internal affairs of the city: a threat most embodied by the activism of the German National Museum. The *Frankish Courier*, a Nuremberg daily, repeatedly published objections to the museum's incursions beyond their "cloister walls" into the "foreign territory" of the city. The paper accused Director Essenwein of being emboldened by the museum's status as a "national institution," rich with financial contributions from home and abroad, to interfere in Nuremberg's municipal affairs.[30]

Beyond fears that the push for preservation violated the city's right to self-government, some Nurembergers also resented the outside interference simply because the walls posed an obstacle to city-planning projects, limited traffic, and were linked by proximity with the cramped, dingy working-class housing found along the city's moats.[31] As tensions between liberals and the growing ranks of Social Democrats rose because of the acute housing crisis faced by Nuremberg's working class, the removal of the city walls provided a project of common interest and appealing symbolism. The Progressive Party had already called for the removal of the city wall in its entirety in 1869, and the magistracy complained about the walls to the Bavarian Interior Ministry in 1870. After the housing crisis of 1872, the project took on new urgency. The land taken up by the sprawling fortifications was prime real estate that could be used for working-class housing. Some of the city's densest and poorest housing could be found along the walls and the question of how the city planned to deal with the growing ranks of the homeless became a point of particular tension between the city magistracy and the Social Democrats. The magistracy's efforts to remove the walls culminated in 1876, when the mayor released a new city plan proposing the replacement of the entire wall circuit with a ring road inspired by Vienna's Ringstrasse.[32]

By the 1880s a growing acknowledgment of the desirability of extensive preservation in Nuremberg matched the cooling of alarmism over the wall's destruction in the national art historical and preservationist community, and the issue slowly dissipated. In 1883, when the subject of Nuremberg's walls came up in the Reichstag, one representative judged that "the unavoidable requirements of hygiene and political economy have been brought into balance with the demands of romanticism for the time being," and the topic was dismissed. In 1887 Bavarian general conservator Wilhelm Heinrich Riehl moved to oppose a compromise plan suggested by the city to partially fill in the moats to form a walkway, using Nuremberg's value as a tourist attraction to justify his opposition. The Nuremberg press represented Riehl's efforts as an attack on the city's autonomy. He dropped his efforts, and the plan was carried out in 1890. In 1888 the Mittelfränkischer Architekten- und Ingenieur-Verein (Middle Franconian Architects and Engineers Society) issued a memorandum both acknowledging the desirability of preserving as much of the walls as possible and asserting that some portions of the walls would need to be removed for the sake of a sensible urban development. The image of Nuremberg as the "treasure chest" of the German nation was one imposed from outside, but over time it came to shape Nurembergers' self-image too. By 1890 Nuremberg had ceased to see the remaining walls as primarily an obstacle to growth and instead viewed them as an opportunity for local promotion. In that year, the city even began considering the restoration of a portion of the wall so recently removed.[33]

In the early years of the empire, the national conflicts over the preservation of Nuremberg's walls became mired in larger debates about the ownership of a shared German past and the divisions—cultural, regional, sectarian, political, and economic—that threatened this unity. Locally, the walls were most valuable to specific interest groups such as like artists and patrician families who had specific reasons to value what the walls represented. By the 1880s there was a growing consensus that Nuremberg's walls were valuable as history and as a tourist attraction but were not to be valued at the expense of the city's modernization. Instead, they became "islands of tradition" within a modernizing city.[34]

CITY WALLS AND HISTORICAL PRESERVATION IN THE GERMAN EMPIRE

As Nurembergers looked to the example of Vienna and its grand ring road in planning their city's defortification, preservationists feared that smaller German cities would in turn look to Nuremberg and tear down their walls and gates until none were left. Munich-based artist Jso Krsnjavi remarked with concern that Regensburg's city government envied Nuremberg's "practical" and unromantic approach to its walls. As Regensburg imitated Nuremberg's destructive policies, Nuremberg, he wrote, wished to become "just another Stuttgart."[35] Fearing for the future of the iconic walled Bavarian town of Nördlingen, an author in the *Zeitschrift für bildende Kunst* (Magazine for the visual arts) fretted that "after the example of Nuremberg, the Nördlinger feel cramped by their Deiniger Gate Tower, and seek its destruction." The desire to tear down walls and gates was, it seemed, contagious. "If it continues like this," the author warned, "soon the last traces of the old glory of the former imperial cities will have disappeared!"[36] An article on the small Bavarian city of Schwabach's decision to remove its fifteenth-century walls in 1874 found evidence of a widespread ill in the small town's choice. "It is well known," the author complained, "that in many of Germany's cities the city walls and towers that have been preserved until today are no longer held in favor by the 'aesthetic sense' of the population." Now, "even Schwabach" had succumbed to this "fashion." The author dismissed as trivial the reasons for removing the walls, asserting that city governments rushed to demolish their old borders as soon as a single resident complained that the walls obstructed his view or that garbage deposited in the walls' shadows gave off a stink. This was the fear of the "frightful leveler" of modernization.[37]

City gates and walls were particularly fraught objects for preservation. Within the context of an individual city's development, as in Paderborn, a preserved wall might indicate a failure to grow and flourish as a city ought to: that the city was stuck in the past. The question was how to strike an appropriate balance between reverence for a superseded but cherished past and the celebration of its supersession in future growth. Civil engineer Reinhard Baumeister captured this paradox when he suggested that preservationists shift their focus from the narrow goal of protecting the physical walls themselves to measures that would encourage the city to flourish in the present as it had in the past those preservationists so cherished.[38] Although Nuremberg was a special case, debates over preservation or demolition were animated by the same question: how to decide when a wall or gate was simply an outgrown and cumbersome relic and when, as a surviving representative of a once omnipresent structure, it was something special worth preserving. In going from the rule to the exception, gates and walls in the latter decades of the nineteenth century raised questions of the relationship between particular local pasts and shared stories of national and regional identity.

Many feared that a boundless "mania" for urban improvement led to gratuitous demolition and the leveling of historical streetscapes that amounted to "vandalism" rather than improvement. In a cartoon printed in the popular illustrated magazine the *Fliegende Blätter* (Flying pages), a crowd of tail-coated bureaucrats, with pickaxes at the ready, stare ominously at one of Munich's most celebrated monuments, the Propyläen, appraising it as simply "yet another" gate to be torn down in the name of progress (see figure 5.3). Their hatred for the symbolism of the city gate, in other words, led them to see all gates as indistinguishable obstacles to development. Because the 1870s were a period of rapid growth in German cities, city walls and gates became particularly powerful symbols of the transformation of the urban landscape. Although most city walls were often no longer located on the urban edge, city walls continued to signal the division between country and the city. Just as the interest in and the delineation of regional differences intensified when unification seemed to threaten their persistence, anxiety about the loss of clear distinctions between the city and the countryside heightened interest in the markers of those obsolete borders.[39]

There were three main frameworks for evaluating the importance of individual city gates—aesthetic, historical, and infrastructural. Professional preservationists more often focused on aesthetic and art historical criteria to determine a specific gate's value. The preservation of city gates for aesthetic reasons as architectural monuments gained importance because of a renewed enthusiasm for the architecture of the German Renaissance.[40] Political discussions of preservation were more likely to emphasize the historical relevance associated with the establishment and geographical expansion of German territory. In arguing for the preservation of one of Düsseldorf's remaining city gates in 1894, a delegate from the city in the Prussian House of Representatives (Haus der Abgeordneten) evoked the importance of this individual gate to the history of the Rhineland and its place in the nation.[41] This city gate, in other words, was not simply a symbol of a long dead system of urban defense but, rather, a monument to the military history of the German nation.

As symbols of defense, city gates evoked past military engagements, and so were especially well suited to become monuments to the victories through which German territory had been both defended and expanded. In Flensburg, a town on the German–Danish border, the Prussian government in Schleswig refused repeated requests for permission to remove the city's sixteenth-century North Gate, because of its artistic and historical value. The city council found these claims of the gate's historical significance laughable. As a fairly typical example of a Renaissance-style gate, why would this structure have been ascribed such historical value by Prussian officials?

5.3 In this satirical cartoon, the object of a city council's drive for "improvement" is not just any old rundown city gate but, rather, the Propyläen in Munich, modeled by celebrated architect Leo von Klenze after the entrance to the Propylaea at the Acropolis and completed a mere eight years earlier in 1862. "Nivellirungs-Vandalismus in Rathhausen," *Fliegende Blätter* 54, no. 1344 (1870): 126, Universitätsbibliothek Heidelberg-Digitalisierung.

But the gate marked the site of military engagements between German and Danish forces in both 1848 and 1864 and opened from the city onto a cemetery with war dead from both these conflicts. As the northern gate of the nation's northernmost cities, situated on the border with Denmark, Flensburg's North Gate signified the 1864 victory in which Schleswig and Holstein had been acquired.[42] Already signifiers of the urban border, city gates represented national frontiers as well.

Another defense of the city gate as monument referred not to the larger history of German territorial development but, rather, to the relationship between the German national character and the urban environment as the incubator of German citizens and values. Making the case for the preservation of several of Cologne's medieval gates after the city lost its status as a fortified place in 1881, an article in a local paper evoked the walls' significance as symbols of the city's legacy of valiantly defending its urban freedoms. The author argues for Cologne's particular importance in connecting Germany's glorious present to a proud past: it "was a city that always knew how to defend its freedom and independence victoriously against besiegers from its walls and fortresses! . . . If the walls of the city have to fall, at least allow the main gates to stand as honorable witnesses of bygone centuries." As in Nuremberg, Cologne's civic visual identity had national significance. Celebration of urban monuments married the national to the local and the urban to the national.[43]

In the 1890s preservation advocates took advantage of Wilhelm II's more energetic attempts to establish an official national culture and argued for the maintenance of city gates as tools for a public education. Jurisdiction over gates remained confused, and battles over preservation could concern each of the ministers of trade, education, culture, and the military precisely because significance could be interpreted in so many different ways. The many paintings produced by Düsseldorf artists of the city's Berger Gate were themselves considered useful tools of public enlightenment, much like the large-scale depictions of heroic past deeds associated with monumental painting. Both the gates and the artistic reproductions of them helped defend against what one representative called the "cynicism" and "internationalism" infecting the nation's aesthetic sensibilities.[44]

Viewed from a different perspective, the evocation of the city's past by preserved walls and gates was not educational or beautiful but, rather, an oppressive reminder of days of urban stasis and uncomfortable conditions. Architect and city planner Josef Stübben, although an advocate for preservation, acknowledged that city dwellers harbored an understandable animosity against city walls and gates as reminders of an unpleasant past. In cities where residents had "suffered so long and hard under the confinement of the cramped city walls . . . a deep aversion against the gates as against the wall prevailed."[45] The resentment of walls as oppressive reminders of a narrow unenlightened past was part and parcel of resistance against preservation campaigns more generally, as local property owners and city councilmen argued against preservation ordinances that they believed obstructed the natural desire for the improvement and modernization of urban space.[46] That the gates had become worn down with neglect in recent decades further intensified the visceral aversion to these markers of an outgrown mode of spatial organization. Speaking to the preservation of gates in Cologne and Bonn, Stübben thought that the best way to heal the old wounds left by the walls was with time: avoid taking rash actions while painful memories of the wall were too fresh; given time, the majority of city dwellers would come to view more objectively the value of city gates as historical monuments. Busso von der Dollen captures the animosity of the battle over Bonn's one remaining medieval gate in the 1890s, as well as the gate's conflicted symbolism, by observing that after its removal in 1898 a memorial plaque commemorating the demolished monument was installed "like one places a monument to a dead tyrant."[47]

Since most Germans agreed that the modern city had outgrown its need for gates, an individual gate had to merit unusual consideration in order to transcend its more pedestrian significance as the relic of outdated systems of taxation and urban control. In other words, to be worth preserving, a city gate had to be invested with a meaning that extended beyond the city in which it was to be found. This was the paradox of city walls and gates as objects of preservation: though profoundly local monuments, their significance stemmed in part from an ability to evoke a transformation of urban space that had simultaneously (at least in theory) reshaped all German cities. Hence, preservation of city gates required negotiating the complicated relationship between the particularities of local histories and landscapes and the commonality of the urban narrative of growth and development. In 1894 the Prussian minister of culture reflected on this delicate balance. Confronted with opposing requests for the preservation and demolition of city gates in two different cities, he declared, to general amusement: "How am I to find a compromise? Perhaps in this: that one gate will stand, and the other will fall."[48]

URBAN GROWTH AS NATIONAL FUTURE

In the same decades that urban walls and gates became ambivalent objects for urban preservation, to be selectively conserved as part of a national historical heritage and mythologized urban past, there emerged a social scientific understanding of German national consolidation and unification that rooted the national German community in the history of "the city," in general, and modeled the growing nation on the growing city. The concurrent celebration of urban borders as cherished symbols of German civic virtues and as oppressive relics of the past from which national unification could free the city were not necessarily contradictory but, rather, represented two sides of the same coin, placing a mythologized spatial narrative of the historical development of the city at the center of German national identity. Walls and gates were important in part because they represented the past but also because, in their presence, they evoked the growth that had overcome them. Throughout the life of the Kaiserreich, the work of economists, historians, geographers, legal scholars, and sociologists provided different theoretical and descriptive frameworks for linking the historical city to the place of the modern city in contemporary German life. In these frameworks, urban flourishing presaged national unification and even expansion.

In 1877, on the eve of Berlin's entry into the "one-million club" of great European metropolises, urban reformer Julius Faucher speculated about the natural limits of the urban growth that had in recent decades seemed so inexhaustible. Even as Berlin attained its one-millionth resident, he doubted whether any city would ever reach the landmark of an eight-digit population: "Who can know, whether instead . . . city and country will merge with one another entirely? Whether instead the network of railway lines will become for the whole nation together like the road network of an integrated city, which will develop from the reciprocal encounter of heretofore divided cities; and whether instead the forests and fields of the whole nation will become like the gardens of this united city? Big cities should not be obliterated from the face of the earth; they must instead grow until they have overgrown the confrontation between city and country." With a dig at Bismarck's oft-quoted declaration that "the big cities must disappear from this earth," Faucher presented a radical vision of national unification based on a model of urban growth in which the united nation was a "united city": "The reciprocal encounter of heretofore divided cities" signified not only the triumph of the urban landscape but also a fusion of regional state with state in the new empire. By foreseeing the collapse of the distinction between city and country, on the one hand, and of city and nation, on the other, Faucher anticipated the de facto erasure of internal German–German borders. The mingling of Dresden's and Leipzig's suburbs with Berlin's would presumably render the distinction between Prussia and Saxony obsolete.[49]

Faucher's vision of the nation as a united city recalled early projects for the economic unification of German lands. In the 1830s Friedrich List's model for a German nation united by a dense transportation network of railways and canals resembled nineteenth-century plans for urban expansions. List called for railway networks to follow a "webbed concentric form" and saw the nation as an expanded city. A well-developed train network, as its advocates anticipated, would bring one city's suburbs into the next, overcoming the traditional divisions not just between city and countryside but between city and city.[50] As an early housing reformer, Faucher fought for a more sensible housing policy in the pages of the *Vierteljahrschrift für Volkswirtschaft und Kulturgeschichte* (Quarterly journal for national economy and cultural history), which he founded in 1863, as a part of the Housing Reform Movement, and as a representative of the Progressive Party in the Prussian House of Representatives. He believed that rising land prices caused by an effective monopoly on a limited housing supply was the major cause of deteriorating housing conditions. Seen by some as a dogmatic Manchesterite, Faucher was an important and controversial figure in the 1860s and 1870s. After his death his analysis of the urban housing market would become very influential in the 1890s as the urban and housing reform movements matured.[51]

Like Sombart, Faucher was, in a way, concerned that German cities remained stunted by the memory of their past, and particularly by the aftereffects of their walls. In his writings on the urban housing problem in Germany, he told the history of the city's spatial organization and growth as a history of its walls and borders. This was an early version of what would become a common narrative explaining the delayed development of Germany's cities compared to their English and French peers. Observing that Germany's city walls remained standing into the nineteenth century more frequently than those of French and English cities, Faucher explained that because German city walls had stayed standing too long, a kind of "illness" hampered these cities' natural development even after their fortifications were removed. When a city had remained walled too long, the mind-set of limited space perpetuated unnecessarily high density development even after the city was liberated from this physical check on its growth. High housing prices, Faucher wrote, formed a new kind of invisible wall around the city center. For the medieval and early modern city, walls provided a necessary protection, but in the recent German past the persistence of those same borders represented the failure of urban land markets and economic development more broadly. The monopoly on real estate that he believed to be the key cause of urban ills, both economic and social, was the psychic and economic legacy of the city wall, and a peculiarity of German urban development.[52]

There was a racial component to Faucher's analysis as well. He believed that low density, rapidly outward-growing cities represented a higher level of cultural development, whereas cities that were towered up (*aufgetürmt*) he associated with an eastern, or Slavic, spatial order, representing a backward mode of social organization. German cities were caught in between the two models of urban development, showing both "English" and "Slavic" tendencies, which together directed their unique path of development. For Faucher, the elimination of all limits on urban growth and settlement within the nation represented the pinnacle of German cultural development because it would mean the victory of a low density model of urban growth associated with cities of the German Northwest like Bremen and Oldenburg over the too fresh memory of the city wall, responsible in large part for the relative backwardness of German city life.[53] Faucher depicted the growth of German cities as the establishment of Germany's proprietorship over its own land—a kind of interior conquest—and expressed the anxiety that a failure to get rid of walls soon enough had somehow stunted German urbanism, with negative consequences for national development. He both hoped and believed that national unification would open up a new era of unhampered urban development.[54]

Faucher's understanding of the city wall as an unnatural limitation on urban growth not superseded soon enough accorded with the view of a number of national economists who wrote in the following decades on city development and the relationship between the city and the nation. For national economists such as Gustav Schmoller and Karl Bücher (as well as younger figures such as Werner Sombart), understanding the growth of the modern city meant coming to terms with the movement of peoples within the borders of the nation. Both Schmoller and Bücher believed that the modern city was a more natural form of settlement than the older walled city, because it was built on the free economic choices of individuals to migrate to urban centers. The modern city could only grow and flourish, however, within the borders of a strong state, which provided many of the services and protections previously provided by the city.[55]

Politically, the nation superseded the city. Bücher argued that the medieval city had grown simply as a result of its walls; all of the other functions that accrued to urban settlements followed from the promise of security offered by urban fortifications. This mode of settlement reached the end of its possible development when there threatened a surplus of labor within the city. After this point, new restrictions placed on migration created a sharper distinction between city and country and a calcification of the urban order that could only be transcended by the transition to a new economic system. Any legal and physical distinctions between the city and the countryside remaining in the present day were remnants of an old economic system that were at odds with the logic of the modern city, which flourished precisely because of the freedom of movement.[56] Modern cities filled functions within the state; they had no independent function of their own except as centers of learning or industry, garrisons, or marketplaces.

The primarily economic and demographic view of the modern city shared by Faucher, Bücher, and Schmoller did not take into account the possibility that it would continue to fulfill any important political role within the state.[57] Insofar as the nation inherited the city's administrative role, these thinkers were not particularly interested in the political relationship between the city and the nation. A city was only as significant as the function it filled. Yet many German liberals around the fin de siècle looked back on the Renaissance as the heyday of enlightened civic culture in German cities and sought to preserve its legacy of communal self-governance.[58] Economists and city administrators looked for a way to reconcile the privileged position of the city within German political life with the dissolution of those spatially defined special privileges that made this position possible. A good example can be found in the introduction to a volume of essays published on the occasion of the 1903 Dresden city exhibition. It was written by another national economist, Robert Wuttke, who argued that the health of the German national idea had always been best embodied by its cities. Germany had enjoyed geopolitical strength only when its cities also flourished. The German burgher's task was to fulfill the higher purpose of the city that was his home in the expanded form of the constitutional nation-state that was the natural inheritor of the self-governing city. These views were characteristic of a new class of enterprising and technocratic mayors, such as Franz Adickes of Frankfurt.[59]

The relationship between city and nation had profound implications not just for the economic development of Germany but also for its political health and governance, as can be seen in the writing and thinking of Hugo Preuß, a legal scholar who would go on to become one of the architects of the Weimar constitution. Preuß took on Bücher's and Schmoller's presentations of the contemporary city as a spatial unit for accomplishing state functions with no independent political significance. In a tract entitled *Staat und Stadt* (State and city) in 1909, Preuß argued that Bücher's mistaken analysis arose from a common misperception of the relationship between city and state. In the past, Preuß wrote, city and state had represented fundamentally opposed principles of political organization. The city was the home of representative self-government, and the state stood for authoritarian principles. The constitutional national government resulted from the combination of what was best about each: the application of urban self-representation in the broader and more "naturally" defined field of the nation. (Natural, he reasoned, because its borders were drawn on the basis of essential differences in language, culture, and geography, rather than the arbitrary borders represented by city walls.) Preuß feared that in Germany the nation's division into a myriad of small states and principalities had disrupted the dialectic of city and nation. Preuß linked Germany's delayed nationhood with the persistence of a division between city and country; a failure of enlightened urban reforms to extend beyond the limited sphere of the small premodern city. German cities, Preuß observed with concern, were "like islands scattered within an agrarian sea." He feared that cities had "not fully escaped from this isolation even today, although not only the stone walls but also the even more divisive barriers of unequal rights ha[d] fallen."[60]

This association of Germany's delayed nationhood with the failure of its cities to grow later played a role in the justification of city expansions through the incorporation of vast swathes of land in the first decades of the twentieth century.[61] For Preuß, the fall of the city wall, along with the mind-set it represented, was the most important prerequisite of the creation of a healthy constitutional state; and for Preuß, as for Faucher, Germany still struggled against the ghosts of its city walls. In the face of this historical failure, Preuß would later seek to construct an alternative political genealogy for the German state, in which political legitimacy sprang foremost from the self-representational governance of the city and Stein's City Code of 1808, rather than Bismarck's Reich, formed the most relevant precedent for the Weimar constitution.[62]

These concepts as expressed by Faucher, Schmoller, Bücher, and Preuß are important for several reasons. For one, they capture the excitement of urban boosterism in the Kaiserreich. For many national economists and ardent free traders, the language of urban boosterism coincided with enthusiastic visions of a centralized nation-state. A second strand of urban boosterism celebrated the very diversity of Germany's urban landscape as a basis for a more diffuse distribution and legitimation of political power in the new empire.[63] These ideas can be traced in the later housing reform and land reform movements. Adolf Damaschke of the Bund Deutscher Bodenreform (Association for German Land Reform) inherited some of Faucher's critique but joined it to a more biological understanding of the city and its place in human settlement.

To ethnographers, statisticians, and geographers the transformation of the city also provided clues to the historical development of the German race and how that development could be mapped onto geographic space. The social biological theory of urbanization began attracting attention in the 1880s and 1890s through the work of Georg Hansen and Otto Ammon.[64] The geographer Friedrich Ratzel, best known for his use (if not invention) of the term *Lebensraum* (living space), gave the city a privileged place in his biogeographical understanding of the relationship between race and space. The place of big cities in Ratzel's writings has often been overlooked, since his theories have been studied primarily in terms of their usefulness to later reactionary and antimodern currents in German theories of nationalism.[65] Although the topic of urban settlement is addressed in a number of his major works, Ratzel wrote two pieces primarily on the subject of cities. One took the form of a travel account of his time in the United States in 1874–1875, *Städte und Culturbilder aus Nordamerika* (Sketches of urban and cultural life in North America), published in 1876. The other was a lecture on the geography of big cities that he gave at the 1903 Dresden City Exhibition as part of a series of presentations on the history and development of the city.

Ratzel believed that the spatial growth of urban centers demonstrated the durability and vitality of a people's connection to the land. He pictured cities as the physical roots linking a *Volk* to its territory. When cities attained a large enough size that it was no longer possible for an enemy to obliterate them, this marked a crucial moment in the history of a people. This was the moment when it became impossible to uproot a people from their home. In this way, Ratzel wrote, "cities have won a higher meaning for the survival of *Völker* and, even more so, for states." Without cities there could be no durable connection between a people and their territory. That is the essential importance of the city to the nation.[66]

In Ratzel's biogeographical understanding of human settlement, no border is ever permanent. Cities, like states, are always in flux, growing or shrinking depending on the state of their health. A successful healthy settlement becomes more rooted over time, the lands of residence and nourishment become inextricably connected, and the big city is born.[67] This process has clear implications for the visible demarcation of urban borders. In the premodern city, the defensive wall once marked the horizon between the city as a home and place of residence and the wider hinterland on which the city depended for its livelihood, but which was too large to be easily secured. This form of spatial organization characterized the early days of a people's conquest of a territory. The defensive border then dissolved as a consequence of the tightening relationship between the city-as-abode and the area on which it depended for its livelihood, until the city no longer had discernible borders. Ratzel identified the blurring of the border between city and countryside as one of the primary distinguishing traits of his age. But even as a growing urban area, the modern city essentially retains defensive importance; it no longer fends off attackers from its walls and bastions but instead provides security by its very size and durability.[68] The historical geography of the city has implications for the spatial development of the future nation.

This argument for the defensive importance of the contemporary city arose from Ratzel's preoccupation with the Darwinian struggle between peoples over space, but his fundamental argument is very like Faucher's or Bücher's in his celebration of the significance of urban growth for national consolidation and expansion. There are also striking parallels between Ratzel's understanding of the relationship between a *Volk* and its cities and certain aspects of German policy toward cities in Polish Prussia beginning in the 1890s. According to Ratzel, the overwhelming importance of the city's defensive aspect resulted from the origin of most cities in an era of conquest, when populations spread thinly over large areas of land and attack was a constant threat.[69] As a people came to possess that land as their own, growth replaced walls. A similar theory, it seems, played a role in the Germanization policy enacted in the 1890s by the Prussian government. This is one of the reasons that it is so important to understand how the spatial relationship between the city and the nation was understood in the German Empire.

POSEN/POZNAŃ, AN URBAN FRONTIER AS NATIONAL FRONTIER

The idea of colonization became part of the framework for new urban planning ideas in the late nineteenth century before the German Empire acquired any overseas colonies.[70] Later, colonization provided new opportunities for applying German urban planning ideas abroad. When Germany acquired colonies in Cameroon, Tanzania, and elsewhere, urban planning was one reference point for administration overseas. Paul Kollmann wrote of the German colonial city of Dar Es Salaam in Tanzania: "Seen from the harbor, the city creates the impression of a German villa colony with its long rows of friendly European-style buildings; to me it seemed as if I found myself in the Havelseen by Berlin." Although, he admitted "we would not on a visit in the hinterland of Berlin see such marvelous palm trees and other tropical plants."[71] In the Chinese colonial city of Qing Dao, administered by the German navy, urban planning likewise served as a technology of colonial administration. As in other European empires, urban planning was one area of technocratic expertise that linked management of working classes in the metropole to the challenges of administering settler colonies abroad.[72]

This was true not just of Germany's overseas colonies but also in the German–Polish borderlands of the Prussian east, in what Kristin Kopp and others have termed Germany's "Wild East." The Prussian settlement commission was buying up land to parcel out in German-owned small farms in Posen/Poznań and West Prussia, the Prussian government was also opening up long-fortified German cities in the east to facilitate German settlement of their Polish hinterlands.[73] Both Danzig/Gdansk and Posen/Poznań were allowed to take down their walls in the 1890s after decades of unsuccessful petitions. The case of Posen/Poznań demonstrates the ways in which urbanization and Germanization campaigns overlapped.

Posen/Poznań had been turned into a heavily fortified garrison city since coming under Prussian control in 1793. Beginning in 1848 the city government and individual residents of the city repeatedly protested against the restrictions on construction in the land outside the fortifications.[74] From the 1870s onward, accelerating urban growth increased dissatisfaction with the fortifications. As traffic increased, the gates became clogged with huge traffic jams. In 1869 Posen/Poznań represented by the city building commissioner, Cäsar Stenzel, was among the founding members of the Association of Fortified Cities of the North German Federation.[75] After decades of unsuccessful petitions, in 1898 the city's desire for liberation from its walls met with a new attitude in Berlin. The interior ministry had come to see the limiting fortifications, once an expression of German power, as an insurmountable setback for Posen/Poznań's "Germanization." The removal of the southern and western Enceinte now seemed a necessary step to prevent Posen/Poznań from becoming "another Prague."[76] Defortification would encourage the residents of Posen to develop a more "German" relationship to their city and to nature.

In 1900 Posen/Poznań's mayor, Richard Witting, declared that "it was among the best and most characteristic traits of the German that he likes to dwell in his own house, however small, that he wants a piece of earth to name his own; some garden land that keeps him in constant connection with nature." The failure of Germanization in the East, Witting argued, resulted from the failure of cities like Posen/Poznań, through the perpetuation of fortifications, to allow this natural relationship to the land to develop: "The Polonisation of Posen in my opinion is largely attributable to the fact that Germans in the province, especially in the Provincial Capital, have so little opportunity to live according to this deeply rooted inclination. Jammed together in narrow, expensive apartments without flowers or trees, without air or light—how ought the German, especially the West- and South-German, feel himself at home?"[77]

Witting, a close associate of Hugo Preuß, made this argument based on an understanding of the history of the city that united nostalgia for a particular pastoralized vision of German urban life with a celebration of urban progress. Witting's vision of the growth of the city as one in which Germans found their patch of garden shared much with visions of the Berlin hinterland as a frontier landscape of small settlements and pioneers from a couple decades before: again, a kind of "settler urbanism," providing a spatial link between rural settlement schemes and urban expansion, reconciling urban and rural ideals. This understanding united nostalgia for a particular pastoralized vision of German urban life with a celebration of urban progress. This is a story that Wilhelm II reinforced when he inaugurated the project in a speech in Posen/Poznań in 1902. His beneficent hope was that the city's "bad old quarters" (*die böse alte Stadtteile*) would soon disappear.[78] In other words, Wilhelm mobilized a familiar story of urban growth to legitimize "inner" colonization and the Germanization of Polish Prussia.

No doubt was left as to the message intended by the demolition: the ring of land opened by defortification was to be used for German cultural institutions like a museum and a grand residence for the emperor, linking urban improvement to a "policy of cultural improvement" much like the ring roads of Hamburg and Vienna. The headquarters of the Royal Settlement Commission was among those administrative buildings constructed in Posen/Poznań in an ostentatious neo-Baroque style evocative of administrative buildings in Berlin in the early years of the twentieth century. Frescoes inside depicted the exploits of German "colonizers" in the East going back to the Teutonic knights and medieval monks. As Gregor Thum puts it, through the architecture of the new Posen, "Prussia's comparatively recent annexation of Polish territory in the eighteenth century was transformed into an allegedly age-old tradition of German expansionism, justified by a civilizing mission at the fringes of empire."[79]

In Posen/Poznań, as in Berlin, the idea of an urban frontier linked urban space to national space. Urban growth and the tearing down of city walls, both literal and figurative, could form the basis not just for national economic unification or political legitimacy based on self-representation but also for expansionism. The linking of Germanization campaigns with defortification and urban expansion harnessed a belief in the inevitability and desirability of urban growth with that of Germanization in Polish lands. As mobilized in Posen/Poznań, the imagery of the growing city unified progressive urban culture with *völkisch* nationalism and helped establish one of those areas of "broad resonance between right and left" that formed the basis for German reform politics between the 1890s and 1910s.[80]

In the German Empire, the city wall was enlisted in a "mobile army of metaphors" by which Germans sought to understand the transformation of their social and physical worlds.[81] In that army of metaphors, the vanishing city wall served as a mercenary, its rhetorical services enlisted in various different narrative campaigns. Those narrative campaigns served to link a sense of time to a sense of place in the cities of the German Empire. The removal or preservation of walls and gates was used to assert the centrality or marginality of a local place. The destruction of city walls could represent an attack on a unified German culture at the heart of the *Kulturnation* in Nuremberg, or its triumph in the contested landscapes of Polish Prussia. At the level of the individual city, urban expansion narratives were often made to align local with global hierarchies of spatial development. In Catholic towns such as Paderborn, modernizing urban space and regulating its borders meant its "Prussianization" from Catholic backwardness. In the Prussian East, German authorities harnessed urban expansion planning as part of a campaign of "Germanization" of Polish space in cities such as Posen/Poznań. The work of Kathryn Cianca shows how, in Eastern Polish towns, the same "civilizing" framework of urbanization, and especially urban expansion planning, was used in analogous ways to facilitate the "Polonization" of Jewish space. These diverse examples demonstrate the flexibility and capaciousness of urban expansion planning as a way of negotiating complex relationships between the local, national, and universal as different scales of experience and understanding. This flexibility arose in part from the city's status both as part of the nation and also as a community that transcended the national in its urban identity and through projects of local cosmopolitanism.[82]

Historical preservation was just one of many discourses linking the nation to the built landscape of "the city" in the German Empire. Social scientists, artists, urban planners, and administrators all used the city's history and spatial development as a framework for interpreting Germany's national consolidation, and even the national territory's need for expansion. Historical preservationists were not distinct in their concern with the past, nor in the way they saw the past inscribed in the physical space of the city, though they may have been uniquely committed to the value of those old stone documents, interpreted and preserved with specialist expertise, as a privileged way of experiencing and connecting to that past. Written historical narratives, social scientific and geobiological theories, artwork, photography, and urban festivities all took the history of the built city and interpreted its relevance anew, connecting Germany's national future to its urban pasts.

Conclusion

"I don't want to stay for always in the city. Yes, for a while more. But when we tire of it, then we move out. Somewhere outside the gates, into the green. There one can regain his strength, resilience, life . . . and our children may go on, and their children, until the city dwellers again become farmers, who will again become city-dwellers."

"Oh, I don't believe that," said Marianne Badekow. "Nowadays everyone would rather live in the city!"

"That is only a transition [*Übergang*], but the city gate is a passageway [*Durchgang*]."

Clara Viebig, *Die vor den Thoren*

Clara Viebig's *Die vor den Thoren* (Those outside the gates), from 1910, is a novel about the city told from the perspective of the suburbs. Opening with the triumphant return of troops from the Franco-Prussian war, it traces the boom and bust of the Gründerzeit through the intertwined lives and losses of several respected families in Tempelhof, a southern suburb of Berlin. Throughout the novel, the characters reflect on how Tempelhof's history and land both seem transformed by contact with Berlin's zealous agents of urbanization. That Tempelhof was once itself a center and an important place—a fortress founded by Templars—no longer seems to matter. From the solipsistic perspective of the city booster, communities outside the gates can have no history of their own, because the city is the driver of historical change. To the building speculator, rich farmland and uncultivated waste are all the same. Farmhouse and hut alike are simply structures to be torn down to make way for rows of apartment buildings. What Viebig's narrative embodies is a process of "wastelanding" that is common to frontier myths: legitimizing growth and development by representing the land as empty and cultureless.[1] The speculators who offer Tempelhof's farmers unthinkable sums of money for their land only do so because for them the land is valuable not as a place with history and soil but as a location that provides proximity to the city. By taking part in those exchanges, the Tempelhofers gain quick wealth but must give up everything they have.

Viebig's nostalgia in *Outside the Gates* is impossible to disentangle from the racism of her popular novels about German settlers in Polish land, but it also provides an interesting contrast. The frontier dramas of *Ostmarkromanen* (Eastern marches novels) such as Viebig's *Das schlafende Heer* (The sleeping army), published in 1903, position Germans as the colonists setting out from center to periphery. In this novel of the urban frontier, Viebig's German heroes are both peripheral and static: encroached upon by the German capital. The constructed myth of German colonialism is inverted: here the inheritors of the Knights Templar are not venturing out but are themselves threatened with displacement. Yet, the conclusion of the book is a hopeful one. A virtuous doctor together with his wife make a new life in Tempelhof outside the city gates. In the book's final passage, the doctor provides an alternative vision of the city edge, in which it is not a place of "transition"—of transforming the countryside into the city—but a place of transit and exchange. He wishes to see the urban edge not in the image of a frontier but in that of a passageway, a road, a city gate. In doing so, he challenges not only the spatial ordering of center and periphery, city and suburb, but also the history of the city as one of unidirectional growth. Instead he offers the persistence of the city gate and the spatial difference it represents, along with a history not of decline or progress but of cycles: "the city dwellers again become farmers, who will again become city dwellers."[2]

This was a radical vision. In an era when people were increasingly on the move, from countryside to city and from old world to new, contemporaries saw the world organized by these journeys into spaces of arrival and departure.[3] In this geography, being a place of departure was tantamount to being part of the past—a past that could be either the location of romantic hopes or a loathed state of stasis. The symbol of the city wall and narratives of its fall played a key role in this geography of arrivals and departures, with powerful twentieth-century reverberations. City growth was enacted in the recapitulation of the fall of city walls and the urban community imagined in the reconstruction and recollection of boundaries long superceded: in the knitting back together the panorama where the wall had once been, in the urban peripheral common, in the embedding of onetime defensive architectures into the "natural" landscapes of the Heimat, in the vexed preservation of walls as memorials, in frontier wildernesses that were both sites of new development and of forgotten pasts. The history of the city as expressed through its borders is a history that necessarily "links past to present dynamically," connected by "lines of genealogical descent, uncanny returns, haunting traces and spectral forces, or nonsynchronous contradictions within an uncanny now."[4] The edge of the city reveals no linear development but is full of dreams of future frontiers and ghosts of walls past.

In English, the word "frontier" can denote either a linear boundary or a border zone or region; it can be either a line of defense or a space of expansion. This ambiguity does not translate easily into German, in which none of the available terms (such as *Grenze* for border, *Grenzland* for border region, or *Rand* for margin or edge) has the same historical and cultural evocativeness that "frontier" has in English. Like Mack Walker's term "hometown," the "frontier" is a concept shaped by an American context and further enriched by a historical myth of American exceptionalism.[5] It calls up images of homesteaders, continental expansion, and colonial conquest with no precise German analogs. When Frederick Jackson Turner's frontier thesis is translated into German, the distinctive word "frontier" is most often preserved, rather than translated (*Frontierthese*). And when German historian Jürgen Osterhammel writes of the global nineteenth century as one of colonial expansion and the "conquest of space" to signify the global processes instantiated in North American westward expansion he too uses the borrowed English word "frontier."[6]

What, then, has it meant to write the history of the German city with the characteristically American idiom of the "frontier"? My use of this term is motivated by a particular argument about the nature of urban growth in the nineteenth century that comes both from my sources and from my analytical approach. Nineteenth-century Germans deployed images of global frontiers to characterize environments of expansion, opportunity, and precarity at the peripheries of German cities. It was not just in fiction that authors such as Gustav Freytag or Clara Viebig offered the suburbs as an alternative to migration to either the Americas or Poland.[7] In the eyes of contemporaries, the provisional market booths of Leipzig's promenades looked like a German "San Francisco."[8] Berlin's boom year shantytowns recalled popular novels about the North American frontier. A persistent consciousness of transatlantic migration shaped plans for and understandings of urban growth. When an Oldenburg native approaching his ninetieth birthday resolved to record his hometown's transformation into a flourishing city in 1874, he framed his story first and foremost within the context of transatlantic migration: "I live alone," he writes in the opening passage. "My children are all in America and my thoughts are often with them."[9] He reflected on how astonished his children would be by the ways in which Oldenburg had become a big city in the past years, but he knew his children would never witness these changes for themselves. He wrote down his memories of Oldenburg's changing borders hoping that they would not be lost between generations now separated by a transatlantic journey. Around 1900 Germans mapped all sorts of global environments onto local landscapes: suburbs called "Little Cameroon" or "New Cameroon" outside Oldenburg, Hamburg, Ulm, and Berlin. Throughout the pages of this book are Germans who connected local landscapes to global processes, and the use of the term "frontier" makes clear that a global consciousness was a key piece of how Germans experienced urban growth.

Osterhammel observes that in the nineteenth century's age of transformation, "the extreme opposite of the city [was] no longer the countryside, the realm of the earth-bound farmer" but, rather, "the frontier." By "frontier," Osterhammel denotes several related things. The frontier is a "movable border of resource development," the outer edge of settlement as it advances into an imagined empty space, and finally, a process of encounter between foreign peoples, one of which must play the role of the invader. The invading settler sees his claim on new ground justified by his right as a conqueror, by the self-evident good done by his transformation of wilderness into productive land, or simply because no one is there to challenge him.[10] Nineteenth-century German urban boosters deployed each of the three justifications for settlement that Osterhammel identifies as characteristic of the frontier. They imagined the city conquering its hinterlands, expanding into essentially empty space, and turning fallow land to productive good. Burgher pioneers struck out into the forest at Leipzig's edge to build their homes and claimed a right to live in suburbs without walls. Oldenburg's city government strove to conquer both moor and suburb as the rightful space of urban expansion. Magazine writers saw Berlin's homeless settlements in the 1870s as urban homesteaders, rejecting the restrictions of the city while embodying its very spirit. In frontier sagas like these, the city conquered a wilderness. In its growth, the city increasingly seemed like a wilderness to be conquered as well. City planners tamed the wilds of spontaneous development at the urban edge with regular roads and centralized infrastructure, and reformers' civilizing projects targeted urban wilds of another kind.[11] In the spatial imagination of nineteenth-century Germans, civilization encountered unsettled wilderness, empty space, and unfamiliar cultures not only in far-flung landscapes of Polish borderlands, the Asian steppe, African colonies, or the American Midwest, but also at the city edge. The nineteenth-century idea of the frontier was a "myth of development."[12] It is in this form—as a structuring myth of development—that the term "frontier" can usefully be applied to urban expansion as well.

The periphery of the city was not, however, a wasteland—or rather, if it was, that wasteland was not an empty or unproductive place. In cities like Leipzig, Oldenburg, and Berlin, the myth of the city's frontier development confronted fundamentally different models for spatial organization at the urban edge. Cities had histories embedded in their regional environments that subverted urban boosters' attempts to think of the city as conquering its hinterland. The ghosts of Goethe and Tacitus were to be found already haunting the primeval forests at the city's edge. Rural communities outside Oldenburg defended their right not just to land but to a spatial system that persisted in distinguishing between the city and the countryside with a wall, appealing to regionally vital technologies of water control as justification. In Berlin, working-class residents used urban peripheral spaces in ways that defied the ideologies of home and nature that dominated the middle-class city.

Through all these changes, the reference point of the city wall—even long after its disappearance—remained a powerful presence in the city's geography, both in the few places where its built traces remained and also in the many more places where they did not. At the end of the nineteenth century, Germans struggled with the implications of the legacy of the walled city for the political and aesthetic cultures of their empire, as it represented both past limitations overcome and cherished traditions of community. As a founding myth, the fall of the city wall had the potential to bridge the opposition between space and place, between civil society and community. The city is the space in which the wall falls again and again but is always still there, the space that is always exceeding its boundaries, changing as it incorporates new people and spaces, the place where the city wall opens to the urban frontier.

Notes

INTRODUCTION

1. Freytag, *Die verlorene Handschrift*, 25 (translations are my own unless otherwise noted); Kopp, *Germany's Wild East*. On Freytag's novel and the uncovering of the German past, see Lonner, "History's Attic." On Tacitus and the German affinity for frontier myths, see Penny, *Kindred by Choice*, 19.

2. Yair Mintzker calculates that in the eighteenth century, there were roughly twelve hundred German fortified cities with populations greater than one thousand. Mintzker, *Defortification of the German City*, 87–88. The cities most likely to be walled were those that had received charters before the thirteenth century, were near a border, and were either a trade center or founded by a ruler or in the shadow of preexisting castles or churches. Tracy, "To Wall or Not to Wall."

3. Friedrichs, *Early Modern City*, 21–22; Obitz, *Geschichten vom Wall*; Papke, *Festung Dresden*; Loeffler, *Geschichte der Festung Ulm*.

4. Walker, *German Home Towns*, 27–31. On the link between cities, states, and territorial fixity, see Maier, *Once within Borders*, 57. Walker's description of the German hometown—though he only applies it to walled cities with between 750 and 10,000 in population—has often been taken as a more general description of uniquely German urban characteristics. See for example, Ribhegge, *Stadt und Nation*, 48. See also Friedrichs, *Early Modern City*, 10–15. On the symbolic and ritual functions of the urban border, see Bauer and Rahn, *Die Grenze*.

5. For an overview of German defortification, see Mintzker, *Defortification of the German City*; Weber, "Stadt und Befestigung." For comparative approaches, see Marion Hilliges on Berlin, Kassel, and Leipzig in "Entfestigung," and Heil, *Von der ländlichen Festungsstadt*. Of the many single city defortification studies, see especially Hahn, *Die Entfestigung der Stadt Saarlouis*; Grobe, *Die Entfestigung Münchens*.

6. Lees and Lees, *Cities and the Making of Modern Europe*, 5; Weber, "Stadt und Befestigung," 301–21.

7. Reulecke, *Geschichte der Urbanisierung*, 17; Muller, "Crumbling Walls." On the demilitarization of fortifications, see especially Lehmann, "Vom Verteidigungsgelände zum Promenadenring," 7.

8. Mintzker, *Defortification of the German City*, 5–7; Walker, *German Home Towns*, 200; Lindemann, *Patriots and Paupers*; Mintzker, *Defortification of the German City*, 89.

9. Aaslestad, "Cities and War," 381. See also Aaslestad, *Place and Politics*. The small number of heavily fortified *Festungsstädte* (fortified cities) were an exception. On these cities, see for example, Tippach, *Koblenz*; Hahn, *Die Entfestigung der Stadt Saarlouis*; Dumbsky, *Die*

deutschen Festungen. On the aftermath of defortification and the ongoing effects of fortification on the social structure of a city, see Schmidt, *Begrenzte Spielräume*.

10. This spatial narrative continues to structure much writing on the European city. See for example Benevolo, *European City*, xv.

11. Bergerson, "History of Neighborliness." The details on walking the walls are omitted from his book on Hildesheim. Bergerson, *Ordinary Germans in Extraordinary Times*. See also Bergerson, "Raum und Zeit."

12. Rose, *Der Bremer Wall*, 38. See also Duntze, *Geschichte der freien Stadt Bremen*, 620; Jackson, *A Sense of Place*.

13. See Koshar, *From Monuments to Traces*; Koshar, *Transient Pasts*; Hagen, *Preservation, Tourism and Nationalism*.

14. Erlin, *Berlin's Forgotten Future*, 24. On historical consciousness around 1800, see also Fritzsche, *Stranded in the Present*; Koselleck, *Futures Past*.

15. Zimmer, *Remaking the Rhythms of Life*.

16. Mintzker, *Defortification of the German City*, 255. See also Tuan, *Space and Place*; Casey, *Fate of Place*.

17. The phrase "urban frontier" has been used before in a few contexts, none directly equivalent to the way it is applied in this book. See Wade, *Urban Frontier*; Merriman, *Margins of City Life*. On the "once-for-all moments: of founding, revolution, liberation" that accompany narratives of progress and growth in the modern social imaginary, see Taylor, *Secular Age*, 716.

18. Roth, *Flight without End*, 100. See also Harvey, *Paris*; Berman, *All That Is Solid*, 17.

19. Schwartz, *Spectacular Realities*; Fritzsche, *Reading Berlin*; Larson, *Constructing and Resisting Modernity*.

20. Wood, *Becoming Metropolitan*, 51; Bryant, "Into an Uncertain Future."

21. Schorske, *Fin-de-Siècle Vienna*, 24, 32. On the city as the natural habitat of the liberal middle classes, Lothar Gall's project takes a very different approach. Gall, *Stadt und Bürgertum*.

22. Applegate, "Senses of Place"; Blackbourn and Retallack, *Localism, Landscape*.

23. Applegate, *Nation of Provincials*. *Heimat*, a notoriously difficult word to translate from German, evokes both the safety and security of one's family "home" and also the romantic national or regional allegiances of one's "homeland." See also Blickle, *Heimat*.

24. Jenkins, *Provincial Modernity*; Prokopovych, *Habsburg Lemberg*.

25. Jerram, *Germany's Other Modernity*; Maderthaner, *Unruly Masses*.

26. Umbach, *German Cities*, chs. 2–4. On the transitional spaces on the urban edge, see Dollen, "Stadtrandphänomene"; Johanek, *Die Stadt und ihr Rand*.

27. Cronon, *Nature's Metropolis*, 23. For a few examples from the genre of city expansion studies, see Gruber-Corr, *Stadterweiterung im Rheinland*; Kier, "Die Stadterweiterungsplanung; Reck, *Die Stadterweiterung Triers*; Pingel, *Stadterweiterung und städtische Behörden*; Fehl and Rodriguez-Lores, *Stadterweiterungen*; Engeli, "Stadterweiterungen"; Rönnebeck, *Stadterweiterung*; Hartog, *Stadterweiterungen*.

28. Brantz and Dümpelmann, *Greening the City*, 2–3; Hennebo, *Entwicklung des Stadtgrüns*. For promenades and parkland, see Clausmeyer-Ewers, *Die Wallanlagen*; Haunfelder and Olliges, *Die Promenade in Münster*; Hans Pritzl, *Mainz im Wandel zu einer offenen Stadt: Der Mainzer Grüngürtel*, Inaugural-Dissertation zur Erlangung des Grades einer Doktors der Technikwissenschaften, vorgelegt der Architektur-Fakultät der Erzherzog Johann Universität, 1993.

29. Jackisch, "Nature of Berlin," 307. See also Ladd, *Urban Planning and Civic Order*, 68, 110; Heil, *Von der ländlichen Festungsstadt*, 10; Tippach, *Koblenz*, 92.

30. Arminius (pseudonym of Countess Adelheid von Dohna-Poninski), *Die Großstädte*; Beta, *Wohl- und Uebelthäter in unseren Großstädten, Deutsche Zeit und Streit-Fragen. Flugschriften zur Kenntniß der Gegenwart V*, eds., Fr. V. Holzendorff and W. Oncken 61 (Berlin: C. G. Lüderitz'sche Verlagsbuchhandlung Carl Habel, 1875); Lees, *Cities, Sin and Social Reform*; Baumeister, *Stadt-Erweiterungen*, 192.

31. Wilson, *German Forest*, 100–101; Blackbourn, *Conquest of Nature*, 28.

32. Cosgrove, "Landscape and *Landschaft*," 68.

33. Koshar, *German Travel Cultures*.

34. Certeau, *Writing of History*, 2; Koselleck, "'Space of Experience' and 'Horizon of Expectation': Two Historical Categories," in Koselleck, *Futures Past*, 255–76; Danto, *Narration and Knowledge*.

1: PICTURING THE CITY

1. Dietrich, *Panorama des Leipziger Rings*, 5.

2. *Der Leipziger Promenadenring zur Biedermeierzeit*. On Leipzig's defortification and the construction of the promenades, see, among others, Lehmann, "Vom Verteidigungsgelände zum Promenadenring," 7–15; Hilliges, "Entfestigung."

3. Clark, *Paris and the Cliché of History*.

4. Koshar, *Transient Pasts*, 1.

5. Dubbini, *Geography of the Gaze*, 75, 186. For overviews of the rise of the panorama, see also Oettermann, *Panorama*; Sternberger, *Panorama of the Nineteenth Century*.

6. Julia Emmrich makes a similar argument in her study of Dortmund: "the fortifications functioned as an ordering structure that fashioned not only an overview, but the opposition between inside and outside" the city. Emmrich, *Wachgeküsst!*, 64.

7. Goethe, *Faust I*, 39.

8. Koch, *Geschichte der Stadt Jena*, 207.

9. *Neue Beschreibung von Leipzig*, 8.

10. Green, *Fatherlands*;. See also Retallack, *Saxony in German History*. On the category of "second cities," see Breuilly and Prothero, "Revolution as Urban Event."

11. Bünz, "Leipzig als landesherrliche Residenz."

12. Beachy, *Soul of Commerce*, 3–4. For a discussion of the relationship between Leipzig's Bürgerkultur and Dresden's Adelskultur, see Blaschke, "Die kursächsische Politik."

13. Beyer, "Leipzigs Auseinandersetzung." On Leipzig as a center of the book trade, see Rosenstrauch, "Leipzig als 'Centralplatz'."

14. Goethe, *Goethe's Autobiography*, 209; Blaschke, "Die kursächsische Politik." On the importance of goods and traders from Asia and Eastern Europe for the Leipzig fairs, see Metscher and Fellmann, *Lipsia und Merkur*, 71–79.

15. Leonhardi, *Geschichte und Beschreibung*, 673–74. New tollhouses for the collection of excise taxes were constructed in the 1710s at the outer gates. Große, *Geschichte der Stadt Leipzig*, 1:361.

16. Kevorkian, "Rise of the Poor," 165 (*Bettelbedrängnis*); City Council of Leipzig, "Instruktion des Leipziger Rats."

17. "Bestimmung des Leipziger Rates über die Behandlung 'verdächtiger' schwangerer Frauen und Mädchen, 14. April 1717," in Bräuer, *Der Leipziger Rat und die Bettler*, 147; Diamant, *Chronik der Juden*, 4; Prasch, *Vertraute Briefe*, 98 (gatekeepeers); "Torhüterinstruktion des Leipziger Rates, um den Zutritt fremder und kranker Personen zur Stadt zu verhindern, 23. Januar 1748" in Bräuer, *Der Leipziger Rat und die Bettler*, 154 (sick persons).

18. *Leipzig: Handbuch für Reisende*, 11. On Leipzig's suburbs, see Czok, *Vorstädte*; Schwela, "Zentrum und Peripherie."

19. Schneider, *Leipzig: Streifzüge*, 190.

20. Krögen, *Freie Bemerkungen*, 62. StadtAL Urk. 12, Nr. 23. See Franz, "Leipzig steht ebenso im Rufe," 86.

21. Franz, "Leipzig steht ebenso im Rufe," 86; Schön, "Die Entstehung des Promenadenrings," 11; Geyler, "Die Promenade," 41.

22. "Auszüge aus Johann Salomon Reimers Leipzigischen Jahrbuchs (1714–1771)," in Wustmann, *Quellen zur Geschichte Leipzigs*, 1:224; StadtAL Urk. 78, Nr. 16–Nr. 18. See also the description of renovations along the fortifications land in Lehmann, "Vom Verteidigungsgelände zum Promenadenring," 9; Große, *Geschichte der Stadt Leipzig*, 1:358–62.

23. Prasch, *Vertraute Briefe*, 158, 8–9; Langermann, *Bemerkungen über Leipzig*, 36–38; [Heidecke], *Tableau von Leipzig*, 191; Krögen, *Freie Bemerkungen*, 72. For a more positive depiction, see Tissot, *Die Geschichte der Stadt Leipzig*, 329.

24. Lehmann, "Vom Verteidigungsgelände zum Promenadenring," 7. Leonhardi included the ward (*Zwinger*) in neither "inner" nor "outer" city, but as a category of its own, in his Leipzig guide published in 1799. Leonhardi, *Geschichte und Beschreibung*, 39–40.

25. Equidius Gotthelf Francke, "Untertänigste Rapport über die Local-Besichtigung der Festung Leipzig," February 6, 1770, Staatsarchiv Amt Leipzig (StaatsAL) 20009 Nr. 4828, f. 99.

26. Fritsch, "Über die ersten nach dem Friedensschlüsse einzuleitende Schritte," 181. See also Herzog, "Kursachsens Städte." On how policies protecting Leipzig's markets were depicted as redounding to the benefit of Saxony as a whole, see Beachy, "Reforming Interregional Commerce."

27. Stadtarchiv Leipzig (StadtAL) Tit. XXXVIII Nr. 19, f. 1; J[ohann] G[ottfried] L[angermann], *Bemerkungen über Leipzig*, 12.

28. StadtAL Tit. XXXVIII (F) Nrs. 19, 23, 24; Staatsarchiv Amt Leipzig (StaatsAL) 20009, Nr. 4828; Franz, "Leipzig steht eben so im Rufe," 87; StadtAL RRA (F) Nr. 194. On the construction of the theater, see also Menninger, "Art and Civic Patronage." Details on the demolition work can be found in StadtAL Tit. XXVII (F) Nr. 2b.

29. Hirschfeld, *Theory of Garden Art*, 184.

30. Hirschfeld, *Theory of Garden Art*, 207. This can be seen in illustrations of Leipzig surrounded by a wall of trees, such as the lithograph by Gustav Taubert of Leipzig seen from the west (1855) in Schneider, *Leipzig: Streifzüge*, 312.

31. Sennett, *Conscience of the Eye*, 78. See also Bernatzky, *Von der mittelalterlichen Stadtbefestigung*, 9, 25; Neumeyer, "Landscape Garden."

32. StaatsAL 20009 Nr. 4828, f. 101; StadtAL Tit. XXXVIII (F) Nr. 19, ff. 8–15 (1770 warning); StadtAL Tit. XXXVIII (F) Nr. 29, ff. 35–37 (Dauthe, complaints).

33. Leonhardi, *Geschichte und Beschreibung*, 165.

34. "Die Gebrüder Fatalis," 396. On the role that the embarrassment caused by this article played in spurring the opening of Leipzig's gates in 1824, see the letter from the Stadtmagistrat Leipzig to the Commun-Repräsententen on October 10, 1823 in StadtAL Tit. XXVI (F) Nr. 2c, 22–25.

35. Blum, *Robert Blum*; StadtAL Tit. XIVc Nr. 60; StadtAL Tit. XXVII (K) Nr. 7, vol. 2, 165. Although the documents do not make clear the motivation for this decision, it came in the wake of the failure of a deputation for the "Verbesserung des bürgerlichen Zustandes der im Königreich Sachsen lebenden Juden," created in 1835 by the Leipzig representative assembly to agree on any recommendations. Reinhold, *Zwischen Aufbruch und Beharrung*, 15.

36. *Leipziger Tageblatt*, May 8, 1824.

37. *Leipziger Tageblatt*, April 17, 1841.

38. Petitions from Friedrich Christian Liederitz, July 28, September 10, 1828, StadtAL Tit. XXVII Nr. 7, vol 1; StadtAL Tit. XXVII Nr. 7, ff. 20–22. The former request was satisfied, but the latter met with unsympathetic ears: the promenade was for the good of the general public, not of individuals.

39. Although the ward was clearly not part of the suburbs, its status as part of the inner city was not self-evident either, at least not in the early nineteenth century. In his 1829 book on Leipzig, for example, Carl Gretschel specifies that his description of the inner city is "mit Inbegriff der Zwinger [including the ward]." Gretschel, *Leipzig und seine Umgebungen*, 63.

40. *Leipzig: Handbuch für Reisende*. One group often mentioned among the inhabitants of the council-owned housing in the ward was the city's town soldiers, a group also well represented among the city's beggars. Bräuer, *Der Leipziger Rat und die Bettler*, 89. There is a list of inhabitants of the Peterszwinger in StadtAL Tit. XXVII (K) Nr. 7, vol. 1 (1811–1830).

41. StadtAL Tit. XXVII (K) Nr. 7, vol. 1, f. 31.

42. StadtAL TitXXVII (K) Nr. 7, vol. 1, ff. 45–46.

43. Semmig, "Das wilde Viertel," 18.

44. StadtAL Tit. XXVII (F) Nr. 7, vol. 2, ff. 44–46.

45. Groß, *Geschichte Sachsens*, 215. See also Schwela, "Zentrum und Peripherie." The population grew from 29,792 in 1756 to 35,230 (1811), 62,374 (1849), and 106,925 (1871). From Keyser, *Deutsches Städtebuch*, 122.

46. "Wohin gehören die Bäume?" *Leipziger Tageblatt und Anzeiger*, January 14, 1844. This article came to the attention of the building commission and is to be found as a clipping in StadtAL Tit. XXVII (K) Nr. 7, vol. 3, 124–25.

47. "Eine Vergrößerung der *innern* Stadt . . ." StadtAL Tit. XXVII (K) Nr. 7, vol. 3 (1840–1850), 41 (original emphasis). Compare to Schilling, *Innere Stadt-Erweiterung*. Geutenbruck notes that the *Bürgerschule* gains in majesty from the shape and raised ground of the old bastion and appeals to the picturesque to defend the preservation of differences in height at the city edge.

48. Briel, *Innere Westvorstadt*, 11.

49. "Müllers Geist über dem Moritzdamme schwebend," *Leipziger Tageblatt und Anzeiger*, April 1, 1857, 1207; "Nun frag' ich alle Guten: was wird da mit den *Buden?*" in "Zur Ausfüllungsfrage," *Leipziger Tageblatt und Anzeiger*, April 1, 1857, 1207 (original emphasis); "Gegenwart und Zukunft des Augustusplatz (Eingesendet)," *Leipziger Tageblatt und Anzeiger*, February 12, 1851, 466.

50. Koshar, "Frightful Leveler."

51. Hocquél, *Leipzig Architektur*, 20.

52. StadtAL Cap. 5, Nr. 5, vol. 1, f. 1. The petition is signed by forty-five residents of Gerber Street and Neu Street, about eighty from the Johannisvorstadt, Sandgasse, and Ulrichsgasse, thirty-eight from Grimmaische Steinweg, Neugasse, Johannisgasse, Quergasse, and Hintergasse, and about sixty-five from the Rannstädter Viertel.

53. StadtAL Tit. XXVI (F), Nr. 1c, ff. 11–12. Their 1829 petition was eventually granted in 1835. The lanterns were to be paid for with contributions both from the city government and from the inhabitants of the neighborhood. But by this point, the neighborhood residents had changed their minds. Having heard a rumor that the suburbs were to be united with the city, they assumed they would be able to get city-quality lighting at a lower cost once this had happened. Attempts to collect money for the lanterns met with difficulty, since it was the home of many poor renters. StadtAL Tit. XXXIX (K) Nr. 25.

54. StadtAL Cap. 5 Nr. 5, vol. 1, ff. 17–38. Later, in 1841, the Friedrichstadt suburb was incorporated as well. Kühn, *Ostvorstadt*, 14; StadtAL Cap 5 Nr. 5, vol. 3.

55. StadtAL Tit. XXXVI (F) Nr. 28; StadtAL Tit. XXIVc (K) Nr. 65, f. 13; StadtAL Tit. XXIVc (K) Nr. 71, 31–32; Kühn, *Ostvorstadt*, 15.

56. *Sächsische Vaterlands-Blätter* 3 (May 30, 1843), 401.

57. I did not see this in the archival material. I did find petitions from the adjacent land-

owners on the inconvenience caused by the outer gates and gatekeepers. See StadtAL Tit. XXXIX (K) Nr. 27, for example.

58. *Sächsische Vaterlands-Blätter* 3 (May 30, 1843), 402. Not everyone found this outer fence so disturbing. As Carl Weidinger described it, in his 1860 city guide: "Of suburbs there can be no discussion. A wooden fence surrounds the entire municipal area and the pleasant gatehouses have been moved out to this border." Weidinger, *Leipzig*, 67.

59. Such complaints also had social implications, since the gatekeepers were poorly paid, and the wagon riders and horsemen who bristled at being slowed down were more likely members of the middle and upper classes.

60. StadtAL Tit. XXXIX (K) Nr. 27, 1–2, 20–22. This particular gate seems to have been cause of the most concern. It apparently was closed earlier than other outer gates. In this case a compromise was reached, and the gate was left half opened all the time, so that horses could pass through but not carriages. In this case restricting passage through the gate was not just a matter of collecting taxes but was also used as a way to control traffic flow. The city council was concerned that too many carriages passing through would kick up too much dust.

61. StadtAL Tit. XXIVc (K) Nr. 71 (1850s); StadtAL Tit. XXXIX (K) Nr. 27, ff. 23–24 (1859).

62. Ludwig, "Die Leipziger Messe," 353. Ludwig, "Die Leipziger Messe," 364–65; StadtAL Tit. XXIVc (K) Nr. 71, f. 142. As the demand for space grew, fair activity increasingly moved outside the city itself. As far as I can tell, no concerted effort was made to remove any remaining walls, which were probably absorbed into private use or allowed to decay. The only documents I was able to find deal with the gatehouses. These were for the most part not removed after 1860 but, rather, were rented out by the city. See for example, StadtAL Cap. 57b Nr. 22 on the Tauchaer Gatehouse after 1861. In response to further development, the border had just been pushed further out in a few places in 1858. StadtAL Tit. XXIVc (K) Nr. 61.

63. *Illustrirte Zeitung*, May 2, 1854, 362, cited in Krüger, *Carl Heine*, 58.

64. Lempa, *Beyond the Gymnasium*, 1–2; Schreber, *Ärztliche Zimmer-Gymnastik*, 10–11; Schreber, "Die Jugendspiele." Schreber did not leave behind any tracts on the allotment gardens to which he gave his name. There is some disagreement about the extent of his importance to the movement. In the absence of clear sources, this disagreement has been aggravated by Schreber's second and much darker legacy as a tyrannical father whose son's psychosis became a subject of study for Sigmund Freud. Gert Gröning, for example, asks why Schrebergärten could not just as well be called Piaget-gärten instead. It is true that Pestalozzi, Rousseau, and Piaget all gave the garden a central place in their pedagogical systems, but Schreber's linking of gardening and exercise was distinctive. Gröning, "Wollen Schweizer Kleingärtner?" 16–17.

65. On the history of the German Schrebergärten movement, see Stein, *Inseln im Häusermeer*.

66. Wilson, *German Forest*, 27, 181, 24.

67. Rollins, *Greener Vision*, 85, 200–203.

68. Ross, "Down with the Walls!"

2: CONQUERING THE WASTELAND

1. Bredehorn, *Eversten*, 41. See also Goerlitz, *Die Landeshauptstadt Oldenburg*, 38–39; Nistal, "Oldenburg wird moderne Hauptstadt," 305.

2. Seggern, *Großstadt wider Willen*. See also Wegmann-Fetsch, *Die Revolution von 1848*, 17; Schmidt, "125 Jahre Eisenbahn im Landkreis Oldenburg," 541–42. On the idea of the "imperial" city and the ways in which urban expansion rested on projects of environmental transformation, see Brechin, *Imperial San Francisco*.

3. Gandy, *Fabric of Space*, 8–9.

4. Blackbourn, *Conquest of Nature*.

5. Lekan, *Imagining the Nation*, 16; Rollins, *Greener Vision of Home*.

6. Blackbourn, *Conquest of Nature*, 313.

7. Schmidt, "Oldenburg in Mittelalter und früher Neuzeit," 14; Hinrichs, *Über Oldenburg*, 33–37. See also the map of the Oldenburg's bioregions, in Eckhardt and Schmidt, *Geschichte des Landes Oldenburg*, map 2 insert.

8. Boy, *Die Stadtlandschaft Oldenburg*, 12–13; Poppe, *Zwischen Ems und Weser*, 13.

9. Boy, *Die Stadtlandschaft Oldenburg*, 5, 14; Schmidt, "Oldenburg in Mittelalter und früher Neuzeit," 391. On moor and marsh and urban fortifications, see Menne, *Die Festungen des Norddeutschen Raumes*.

10. Schmidt, "Oldenburg in Mittelalter und früher Neuzeit," 19; Gilly, *Festung und Garnison Oldenburg*, 107. Oldenburg already possessed simple wooden and earthen fortifications when it first appeared in the written record in 1108. It gained a ring wall when it first became the site of the castle of the House of Oldenburg later in the century. Wilhelm Gilly de Montaut, *Festung und Garnison Oldenburg*, 21.

11. From the 1520s to 1615, Oldenburg was a fortress in the Italian style. In 1529 Count Anton I extended the fortifications between the Hunte and the Haaren. Starting in the early seventeenth century these fortifications were updated and expanded in the Dutch style. Under Danish rule after 1689, they reached their greatest extent in the city's brief career as the "Danish Royal Fortress Oldenburg." Gilly, *Festung und Garnison Oldenburg*, 19.

12. It was shared with the duke of Holstein-Gottorp until 1676. Schmidt, "Oldenburg in Mittelalter und früher Neuzeit," 332.

13. Krüger, "Wandel des Stadtbildes," 48, 76–81; Boy, *Die Stadtlandschaft Oldenburg*, 15–16. See also Limann, "Hydrographie der Stadt Oldenburg." Limann provides a superb series of maps demonstrating the changes to Oldenburg's waterscape over the nineteenth century. See also Gaumer, "Leben in Oldenburg"; Meyer, *Oldenburg und das Wasser*.

14. Schmidt, "Oldenburg in Mittelalter und früher Neuzeit," 391; Krüger, "Wandel des Stadtbildes," 60.

15. Kersten Krüger explores in detail the successive plans (only a few of which were ever

realized) for the expansion of Oldenburg's fortifications under Danish rule. These plans combined neo-Italian, neo-Dutch, and French models, relying especially on Vauban and the Dutch fortifications specialist Menno van Coehoorn. The latter, in keeping with the Dutch landscape, relied more heavily on water fortifications. Krüger points out that, although the planners strove for the ideal of a circular arrangement of equally sized and evenly spaced bastions, Oldenburg's topography allowed this style of semicircular rampart only in the north and the south. Krüger, "Wandel des Stadtbildes," 52–53, 55.

16. The later ideal of the "star-shaped" city would echo the contours of star-shaped fortification systems. Gilly, *Festung und Garnison Oldenburg*, 21.

17. "Vorschläge," Count [Lynar] to Minister Bernsdorf Oldenburg, February 11, 1764, Best. 71-3 Nr. 63, 20. Niedersächsisches Landesarchiv Oldenburg (Lower Saxon State Archive, Oldenburg; hereafter abbreviated as NLA OL). For all archival documents, folio or document numbers are included as marked in the document series. For full document series titles, see bibliography.

18. Mayor of Oldenburg, May 9, 1764. NLA OL Best. 71-3 Nr. 63, 61–62, 65 (quotes); NLA OL Best. 262-1Ab Nr. 6163; NLA OL Best. 20-33c Nr. 6.

19. On Peter's reign as an enlightened absolutist, see Schmidt, "Herzog Peter Friedrich Ludwig von Oldenburg"; Schaer and Eckhardt, "Herzogtum und Großherzogtum Oldenburg."

20. "Aufgabe an die Kammer, wegen der mit dem Walle und der Brücke am Harenthor vorzunehmenden Veränderungen, Eutin," February 19, 1796, NLA OL Best. 31-4 41 Nr. 14, 86–87; Hinrichs, "Oldenburg in der Zeit Herzog Peter Friedrich Ludwigs," 559–61; "Aufgabe an die Kammer, wegen der vorzunehmenden Abtragung des Walles vom Heil. Geist Thore bis zum Gefangen Hause," October 18, 1800, NLA OL Best. 31-4 42 Nr. 10.

21. Gäßler, *Klassizismus*, 23. The road leading out from the Haaren Gate was paved in 1782. NLA OL Best. 71-3 Nr. 68, 7–13. In the 1790s, the road south of the palace, called the Damm, was plastered and a number of houses built between branches of the river. The city also made improvements to Hunte Street with the construction of eight large private residences. The street leading out of Eversten Gate was paved and improved in 1807–1808. See Gäßler, *Klassizismus*, 14, 22; NLA OL Best. 71-4 Nr. 27.

22. NLA OL Best. 31-4 41 Nr. 10. See also Gäßler, *Klassizismus*, 19; NLA OL Best. 262-1A Nr. 2185. Among cities in the region, only Münster could boast anything similar. Gruner, *Meine Wallfahrt zur Ruhe*, 236. On Münster's fortifications and promenades, see Haunfelder and Olliges, *Die Promenade in Münster*. For a representative expression of this idea about the relationship between gardens and the surrounding landscape, see "Nachricht vom Oldenburgischen Schloß-Garten," *Oldenbürgische Blätter*, December 29, 1817, 625.

23. Limann, "Hydrographie der Stadt Oldenburg," 143–44; Pollak, *Cities at War*, 2.

24. See, for example, Spieske, *Erinnerungen eines alten Oldenburgers*.

25. See, for example, the discussion over the city's purchase of the former fortifications land between 1906 and 1910 in NLA OL Best. 262-1A Nr. 2241b.

26. Poppe, *Zwischen Ems und Weser*, 19; also Bucholtz, *Aus dem Oldenburger Lande*, 62–63.

27. In 1852 the desire that the number of paths into the city be increased led to the construction of a new street through the old fortifications land and the removal of a small piece of the rampart next to the Staulinie. NLA OL Best. 31-13 66, Nr. 48c, 8–9. The Oldenburg almanac listed the fees and closing times for the gates each year from 1815 on. See for example *Oldenburgischer Staats-Kalender auf das Jahre Christi 1824* (Oldenburg: Schulze, 1824), 288.

28. Lampe, "Wirtschaft und Gesellschaft Oldenburgs," 21; Spieske, *Erinnerungen eines alten Oldenburgers*, 3–4. Another example of institutions associated with criminality and poverty at the urban border is the "Sklavenloch" to the south. Bredehorn, *Eversten*, 462–65.

29. In 1823 an extra man was added at Eversten Gate to handle the growing number of revelers at night. "Bericht des Magistrats der Stadt Oldenburg betreffend die Erhebung des Sperrgeldes am Eversten Thore," October 30, 1823, NLA OL Best. 262-1A Nr. 3721, 14. Hunte Street and the Middle Damm were incorporated in 1808, stretches of the Stau and Garten Street in 1833. On urban expansions, see Braun and Neumann, *Die Oldenburger Neustadtquartiere*.

30. City magistracy to grand ducal government, March 14, 1836, NLA OL Best. 262-1A Nr. 3721. Additional petitions were made by Oldenburg's suburban dwellers in 1836 and 1840, both refused on similar grounds.

31. Bucholtz, *Aus dem Oldenburger Lande*, 42. The most significant attack occurred on the evening of September 2, 1835, when one hundred or so journeymen gathered along the old ramparts by the Heiligengeist and Haaren Gates to shout insults and throw stones at the gate watchmen. Captain Lehmann to City Commandant, September 3, 1835, NLA OL Best. 262-1A Nr. 3301.

32. "Die Thorsperre in Oldenburg," *Neue Blätter für Stadt und Land* 63 (August 9, 1843): 304. At the same time, some Oldenburgers were concerned about how their city and land were seen in neighboring cities. See *Abwehr eines Oldenburgers gegen die Verunglimpfungen seiner Vaterstadt durch Nummer 21. und 22 des diesjährigen Mindener Sonntagsblattes, das aus dem Dunklen ans Licht gezogen ist durch den Ritter vom weißen Torf* (Oldenburg: Verlag von Carl Sonnenberg, 1843).

33. "Die Thorsperre in Oldenburg," *Neue Blätter für Stadt und Land* 63 (August 9, 1843): 304. See also "Die Thorsperre," *Neue Blätter für Stadt und Land* 3, no. 75 (September 17, 1845): 329, 331. In 1836, the city's midwives petitioned for, and were granted, a special exception from the gate tax. NLA OL Best. 262-1A Nr. 3721, 69; NLA OL Best. 262-1A Nr. 2074.

34. "Feier des sechsten Januar 1845 [500 Jahre Stadt Oldenburg]: [nebst] Fest-Programm" (Oldenburg, ca. 1845). See also reprint as "Jubelfeier Oldenburg," *Oldenburgische Blätter* 2 (January 14, 1845): 9–14.

35. "Feier des sechsten Januar 1845."

36. "Entwurf zu der vor den Bürgerversammlung an die Stimmberechtigten zu erlassenden Bekanntmachung," August 31, 1845, NLA OL Best. 262-1A Nr. 3722, document 15. Because the grand ducal government required the city to open their proposal to a general town meeting, the abolition of the gate tax did not in the end occur in time for the quincentennial anniversary but instead followed the next January.

37. Strackerjan, *Von Land und Leuten*, 169, 163.

38. "Oldenburger Zustände," *Der Oldenburgische Volksfreund* 4, no. 34 (April 28, 1852): 135; also Jung and Wundram, *Landschaftsgeschichte des Oldenburger Landes*, 23.

39. "Oldenburger Zustände," *Der Oldenburgische Volksfreund* 4, no. 35 (May 1, 1852): 140. On floods, see Blackbourn, *Conquest of Nature*, 131–32. For a description of worsening flooding in Oldenburg's hinterland, see Meyer, "Naturräumliche Bedingungen," 16.

40. Dierßen, "Hochmoore aus biozönotischer und ökosystemarer Sicht."

41. Lange, *Lebenserinnerungen*, 13.

42. Blackbourn, *Conquest of Nature*, 147. On the practice of moor burning, see Jones, "Fixing Prussia's Peripheries," 204–27.

43. Berg, Meyer, and Steitz, *Moderne Zeiten*, 22, 33.

44. Such as the settlements Friedrichsfehn (1852), Moslesfehn (1855), and Elisabethfehn (1862). See Blackbourn, *Conquest of Nature*, 152–53; Jung and Wundram, *Landschaftsgeschichte des Oldenburger Landes*, 33–34.

45. Moor, Shaw-Taylor, and Warde, "Comparing the Historical Commons," 27; Gudermann, "Conviction and Constraint," esp. 37–39.

46. Starklof, *Moor-Kanäle*, 20, quoted in Blackbourn, *Conquest of Nature*, 151; Blackbourn, *Conquest of Nature*, 149–54; Starklof, *Moor-Kanäle*, 6.

47. Blackbourn, *Conquest of Nature*, 134; Wein, *Stadt wider Willen*.

48. Blackbourn, *Conquest of Nature*, 142.

49. Blackbourn, *Conquest of Nature*, 131.

50. Gudermann, *Morastwelt und Paradies*, 1–3.

51. On Lasius's influence on Oldenburg's development, see Neumann, *Stadtplanung und Wohnhausbau*, 14–15; Deuter, *Oldenburg*, 146–52.

52. Lasius, *Blicke in der Stadt Oldenburg*, 3, 19; also Neumann, *Stadtplanung und Wohnhausbau*, 14.

53. Lasius, *Blicke in der Stadt Oldenburg*, 18; Deuter, *Oldenburg*, 147–48. On the Theaterwall development, see Braun and Neumann, *Die Oldenburger Neustadtquartiere*, 43.

54. Lasius, *Oldenburg zur Zeit*, 4; Starklof, *Moor-Kanäle*, 6.

55. City council and city magistracy meeting, October 17, 1846, NLA OL 262-1A Nr. 1015.

56. City magistracy to grand ducal government, "Bericht vom 30. September 1847 betreffend die anderweite Begrenzung der Stadt und Vorstädte," NLA OL Best. 262-1A Nr. 1015, 10. The city's economy was highly dependent on the state. In 1855, of the city's 10,500 inhabitants 1,800 were employed in the civil service. Parisius, *Vom Groll der 'kleinen Leute'*, 22.

57. City magistracy to grand ducal government, "Bericht vom 30. September 1847 betreffend die anderweite Begrenzung der Stadt und Vorstädte," NLA OL Best. 262-1A Nr. 1015, 10–12.

58. County of Oldenburg, "Bericht vom 18. Mai 1849 betreffend die Vereinigung des äußern Dammes mit der Stadt Oldenburg," NLA OL Best. 262-1A Nr. 1015.

59. Johann Friedrich Mutzenbecher, Grand ducal government to city magistracy, July 31,

1846, NLA OL Best. 262-1A Nr. 1074, document 19. See summary in *Darstellung des von der Stadtgemeinde Oldenburg*, 8–9. On the octroi, see also Schmedes, *Das Octroi in Oldenburg*.

60. "Die Erweiterung der Grenzen der Stadt Oldenburg," *Neue Blätter für Stadt und Land* 64 (August 11, 1847): 271.

61. "Die Erweiterung der Grenzen der Stadt Oldenburg," *Neue Blätter für Stadt und Land* 71 (September 4, 1847): 301.

62. City magistracy, March 22, 1849, NLA OL Best. 262-1A Nr. 1015.

63. Grand ducal government to city magistracy, October 8, 1849, NLA OL Best. 262-1A Nr. 1074 (emphasis in the original). The industry of the entire state of Oldenburg in 1850 consisted of only around forty small factories, among which the largest were six tobacco factories, six glove factories, one iron foundry, and one ship builder. Reinders-Düselder, "Oldenburg im 19. Jahrhundert," 127.

64. On May 13–21, 1851, four meetings were held by the city magistracy for house owners outside the Haaren and Heiligengeist Gates, with a total vote of 193 against and 93 for incorporation into the city. NLA OL 262-1A Nr. 1015.

65. "Erklärung des Ausschusses der Stadtgebiets-Gemeinde betreffend die beabsichtigte Vergrößerung der Stadt durch einen Theil des Stadtgebiets," to the city magistracy, October 12, 1851.

66. Schulze, *Oldenburgs Wirtschaft*, 22, 27; Meyer, "Naturräumliche Bedingungen," 15. On Osternburg's development into an industrial center see Meyer, "Das Industrie- und Arbeiterviertel."

67. Meyer, "Naturräumliche Bedingungen," 15, 20.

68. "Die städtebauliche Entwicklung Osternburgs," in Schachtschneider, *Osternburg*, 134; Meyer, "Das Industrie- und Arbeiterviertel," 89.

69. Berg, Meyer, and Steitz, *Moderne Zeiten*, 136.

70. J. E. Barnstedt, introduction to Oldenburg County, *Oldenburger Actenstücke*, 3.

71. "Gutachten der Osternburger Bevollmächtigten," 6, 7. These areas were in Donnerschweer Sielacht, Drielakermoor, and Osternburg.

72. "Gutachten der Osternburger Bevollmächtigten," 15.

73. "Gutachten der Osternburger Bevollmächtigten," 7–8.

74. Johann Gottfried Hoche, *Reise durch Osnabrück und Niedermünster in das Saterland, Ostfriesland und Groningen* (Bremen, 1800; repr. Leer, 1977), 96, quoted in Berg and Hoffmann, "Das Duis-Album."

75. "Gutachten der Osternburger Bevollmächtigten," 8–9. In 1853 the ditch on the border of the city by Eversten Gate, purchased in 1836 to demarcate the urban edge, was deemed to have lost significance, because of the lifting of the gate closure and the imminent border changes, and so the city offered it to the adjacent landowner. City magistracy to grand ducal government, "Bericht vom 9. April 1853 betreffend den Stadtgraben beim vormals Ritterschen, jetzt Overbeckschen Fischteich," NLA OL Best. 262-1A Nr. 1015.

76. "Gutachten der Osternburger Bevollmächtigten," 17.

77. Osternburg Gemeinde Vorstand to city magistracy, April 26, 1856, NLA OL Best. 262-1A Nr. 1074, Nebenakte A, document 1.

78. Peter Sahlins, in a study of the French/Spanish frontier, explores how the border emerged out of the interaction between local disputes and interests on the one hand and the national project of building a territorialized state on the other. A related set of insights can be brought to bear on the story of the urban border. Both those who dwelled on Oldenburg's urban periphery and the state and city governments strategically deployed understandings of what was natural to and characteristic of the modern city. Sahlins, *Boundaries*, ch. 4.

79. Lekan, *Imagining the Nation*, 8. See also Di Palma, *Wasteland*; Rollins, *Greener Vision of Home*.

80. Heinrich, "Gemalte Heimat"; Küster, "Die Malerei der Moore"; Gäßler, "Wenn Künstler heute ins Moor gehen." On urban preservation, see Koshar, "Frightful Leveler."

81. Such as Starklof, *Moor-Colonien*, 85.

82. "Moorbildung aus Muffrika," in *Die Gartenlaube* 49 (1867): 774–76; Gäßler, "Der Maler und Graphiker Bernhard Winter." On Muffrika as an epithet used to describe moorlands where burning practices were employed, see Jones, "Fixing Prussia's Peripheries," 215.

83. Ohrt, *Die Großherzoglichen Gärten*, 87; Diekmann-Dröge, "Niederdeutsche Heimatliteratur."

84. Poppe, *Zwischen Ems und Weser*, 11.

85. Jaspers, "Oldenburger Welt," 254.

86. For other examples of this kind of romanticization of the city wall in the region, see Rose, *Der Bremer Wall*; Kraus and Weiss, *Das Stadt-Thor in Bild und Wort*.

87. Lange, *Lebenserinnerungen*, 14.

88. Di Palma, *Wasteland*, 2–3.

89. Hinrichs, *Über Oldenburg*, 3–4, 8.

90. See Jung and Wundram, *Landschaftsgeschichte des Oldenburger Landes*, on the emergence of a movement to preserve Oldenburg's unique landscape beginning in the late nineteenth century.

91. Lekan, *Imagining the Nation*, 16.

92. Theodor Goerlitz, *Die Landeshauptstadt Oldenburg*, 5–6, 38–39; Friedl, *Biographisches Handbuch*, 242–44.

93. Goerlitz, *Die Landeshauptstadt Oldenburg*, 6.

94. Schmidt, "Eversten," 10–12.

95. Goerlitz, *Die Landeshauptstadt Oldenburg*, 38.

3: Taxing the Urban Border

1. Vincke quoted in Haase, *Die Entstehung der westfälischen Städte*, 11:197; "Allerhöchste Kabinettsorder, die Erhaltung der Stadtmauern betreffend," June 20, 1830, *Gesetz Sammlung für die Königlichen Preußischen Staaten*, Nr. 15, 113.

2. Mintzker, *Defortification of the German City*, 203–4, 187–88.

3. Grobe, *Die Entfestigung Münchens*, 15; Scott, *Seeing like a State*, ch. 2; Maier, *Once within Borders*.

4. On urban competitiveness as an aspect of modernization, see Zimmer, *Remaking the Rhythms of Life*, 15–16.

5. Siegert, *Steuerpolitik und Gesellschaft*, 84; Spoerer, *Steuerlast*, 68.

6. Haase, *Die Entstehung der westfälischen Städte*, 11:196–97; Gray, "Prussia in Transition," 86. Stein's city code (*Städteordnung*), intended as a first step toward eliminating the legal and organizational distinctions between urban and rural localities, was not implemented in Westphalia in 1808. Here, and in the Rhine Provinces, the existing laws (meaning the French system) were initially preserved.

7. Philippi, *Sammlung sämmtlicher neuer Preußischer Gesetze*, 440–41.

8. Leesch, *Geschichte der Steuerverfassung*, 437; Schönbach, "Preußische Verwaltung," 52.

9. "Bericht des Königlichen Staatsministeriums vom 31. Januar 1820," in Dieterici, *Zur Geschichte der Steuer-reform*, 245–46.

10. Weydemeyer, "Ueber Steuern in Preußen," *Das westphälische Dampfboot. Eine Monatsschrift* 3, 702. Berlin's central slaughterhouse was not completed until 1881. This was relatively late compared to other major capitals. Brantz, "Slaughter in the City," 7.

11. Brantz, "Slaughter in the City," 226. For complaints about this situation, see Landesarchiv Berlin A Pr Br Rep. 030 Nr. 622, 14554.

12. Leesch, *Geschichte der Steuerverfassung*, 438. Reinick, "Resultate der Mahl- und Schlachtsteuer," 222. See also Villaume, *Handbuch der preußischen Steuer*, 7.

13. Mettele, *Bürgertum in Köln*, 271; Schönbach, "Preußische Verwaltung," 52; Keinemann, *Von den Freiheitskriegen*, 252; Hahn, *Altständisches Bürgertum*; Hahn, "Zunftproteste gegen den modernen Steuerstaat"; Reinick, "Resultate der Mahl- und Schlachtsteuer," 217.

14. Schönbach, "Preußische Verwaltung," 53; Leesch, *Geschichte der Steuerverfassung*, 438.

15. The term *Douanenkette* is a reference to the French customs tax system. Weydemeyer, "Ueber Steuern in Preußen," 701.

16. Blackbourn, *History of Germany*, 72. On the economic costs of the pre-Napoleonic customs system, see Spaulding, "Revolutionary France and the Transformation of the Rhine:" 203–26.

17. Prince-Smith, "On Disadvantages to Industry through Increasing Import Tariffs," 4.

18. F. v. Holtzendorff-Vietmannsdorf (1844) quoted in Siegert, *Steuerpolitik und Gesellschaft*, 63:153.

19. Leesch, *Geschichte der Steuerverfassung*, 438–39.

20. Maron, "Vom Ende des Fürstbistums," 12; Spoerer, *Steuerlast*, 68.

21. Schemionek, *Die Mahl- und Schlachtsteuer*, 5 (pamphleteer); Braun, "Taxation, Sociopolitical Structure, and State-Building," 321 (foodstuffs); Spoerer, *Steuerlast*, 68.

22. See Hohmann, *Bauten des Historismus*, 31.

23. Anton Schneider to the Finance Minister, September 5, 1833, and reply, December 3,

1834. GStA PK I. HA Rep. 151 Finanzministerium III Nr. 2050. A number of similar cases can be found in GStA PK I. HA Rep. 151 Finanzministerium III Nr. 2051.

24. A carpenter in Greifswald, for example, petitioned the Finance Ministry in 1831 for permission to put a window in his workshop wall, which so happened to also be the city wall. Other similar cases are included in the document series from an innkeeper in Neu Ruppin and a Stargard coffeehouse owner, among others. GStA PK I. HA Rep. 151 Finanzministerium III Nr. 2050.

25. The provincial government in Koblenz reported to the Finance Ministry in Berlin that Rau had a "raw rebellious" character in a November 1823 report. GStA PK I. HA Rep. 151 Finanzministerium III Nr. 2050.

26. As in the city of Wetzlar. These petitions are collected in GStA PK I. HA Rep. 151 Finanzministerium III Nr. 2050, Acta betreffend den Durchbruch von Pforten und Oefnungen durch die Stadtmauern, vols. 1–2, 1820–1877. They are not organized by region, so the following analysis draws on petitions from across Prussia.

27. See, for example, the appeals of a coffeehouse owner in Stargard in 1821, or a shooting club owner in Schwedt (Oder) in 1862. GStA PK I. HA Rep. 151 Finanzministerium III Nr. 2050; "Etablissement erfreut sich eines zahlreichen Besuches aus Personen der höchsten Stände," GStA PK I. HA Rep. 90A Nr. 1800.

28. These laws are referred to in the petitions in GStA PK I. HA Rep. 151 Finanzministerium III Nr. 2050; *Breslau Amtsblatt*, Nr 361 (1818), 571.

29. " . . . die willkürlich Zerstörung der Mauern den Stadt-Kommune . . ." GStA PK I. HA Rep. 90A, Nr. 1799.

30. " . . . bloßer Liebe zur Veränderung und Verschönerung." GStA PK I. HA Rep. 90A, Nr. 1799.

31. *Allgemeines Landrecht für die preußischen Staaten*, § 33 Tit. 8 Th 1, 1794.

32. "Allerhöchste Kabinettsorder, die Erhaltung der Stadtmauern betreffend," June 20, 1830, *Gesetz Sammlung für die Königlichen Preußischen Staaten*, Nr. 15, 113.

33. See, for example, the case of Soest in 1839. GStA. I. HA Rep. 151 Finanzministerium III Nr. 2159.

34. GStA. I. HA Rep. 151 Finanzministerium III Nr. 2159. Another example is Merseburg. In 1825, tax officials accused Merseburg of opening new city gates and removing a piece of their city wall without permission. GStA PK I. HA Rep. 151 Finanzministerium III Nr. 2050.

35. This number is based on complaints recorded in the Finance Ministry's records from the provincial tax director in Münster. GStA PK I. HA Rep. 151 Finanzministerium III Nr. 2158.

36. See Leesch, *Geschichte der Steuerverfassung*, 438.

37. Letter from the Prussian administration in Münster to Finance Minister Klewitz, May 22, 1820. GStA PK I. HA Rep. 151 Finanzministerium III Nr. 2158. These were the old outer walls and gates of the city. Münster is an interesting case, because it was the site of a much

admired early defortification and promenade project. See Haunfelder and Olliges, *Die Promenade in Münster*, 3, 37. See also Bernatzky, *Von der mittelalterlichen Stadtbefestigung*, 15.

38. GStA PK I. HA Rep. 151 Finanzministerium III Nr. 2158.

39. Improvements intended to stem the flow of smuggled goods were completed in 1825. GStA PK I. HA Rep. 151 Finanzministerium III Nr. 2158.

40. GStA PK I. HA Rep. 151 Finanzministerium III Nr. 2158.

41. Beck, *Origins of the Authoritarian Welfare State*. This was the case in Oldenburg (see chapter 2).

42. "Die Bekanntmachung des Oberpräsident von Westphalen über die gegenwärtigen Noth," *Das westphälische Dampfboot* 3 (January 1847): 55.

43. City of Paderborn, *Bericht über die Verwaltung . . . , 1880/81*, 40.

44. Fuchs, *Paderborn*, 15; Maron, "Vom Ende des Fürstbistums," 7.

45. Fuchs, *Paderborn*, 12.

46. Fuchs, *Paderborn*, 23.

47. GStA PK I. HA Rep. 151 Finanzministerium III Nr. 2158.

48. Gruner, *Meine Wallfahrt zur Ruhe*, 99.

49. Maron, "Vom Ende des Fürstbistums," 74–75, 77; also Heggen, *Paderborn unter preußischer Herrschaft*.

50. Keinemann, *Von den Freiheitskriegen*, 177, 239–44.

51. Landesarchiv Nordrhein-Westfalen Abteilung Ostwestfalen-Lippe, Staats- und Personenstandsarchiv Detmold (North Rhine–Westphalia State Archive in Detmold; hereafter cited as NWStADT), Department Minden, M1-I-K Nr. 2295.

52. NWStADT M1-I-K Nr. 2295 and GStA PK I. HA Rep. 151 Finanzministerium III Nr. 2050.

53. Letter from Finance Ministry, November 22, 1824. GStA PK I. HA Rep. 151 Finanzministerium III Nr. 2158.

54. GStA PK I. HA Rep. 151 Finanzministerium III Nr. 2158. See also Stadtarchiv Paderborn (Paderborn City Archive; hereafter cited as SAPad) A851. I found no evidence that this precedent ever caused difficulty elsewhere. On the other hand, it also seems that Prussian officials were less insistent that other cities pay for the upkeep of their own walls.

55. Uhlenhuth, "Die Paderborner Stadtmauern," 6.

56. Maron, "Vom Ende des Fürstbistums," 21, 23, 40, 67.

57. Sperber, "Competing Counterrevolutions," 51, also 54.

58. NWStADT M1-I-K Nr. 2296; GStA PK I. HA Rep. 151 Finanzministerium III Nr. 1401; Maron, "Vom Ende des Fürstbistums," 23. For accounts of how cholera scares played a role in concerns about the walls and trash deposited along them, see especially the documents on the years 1830–1851 in SAPad A858.

59. From a letter from Provincial Tax Director Göring (Münster) to General Director of Taxes Herr von Pommer Esche (Berlin), June 11, 1851, NWStADT M1-I-K Nr. 2296.

60. NWStADT M1-I-K Nr. 2296.

61. GStA PK I. HA Rep151 Finanzministerium III Nr. 1401.

62. Letter from Friedrich Wilhelm to the Ministers of the Interior, Finance, and War, June 30, 1853, GStA PK I. HA Rep. 151 Nr. 1401.

63. SAPad A842. See also Uhlenhuth, "Die Paderborner Stadtmauern," 13; Beck, *Origins of the Authoritarian Welfare State*, 134. Paderborn was in the Regierungsbezirk Minden within the Province of Westphalia.

64. Letter from Landrat Grasso to Stadtdirektor zur Mühlen, June 27, 1856, SAPad A842; Letter from Stadtdirektor zur Mühlen to Landrat Grasso, June 28, 1856, SAPad A842; Letter from the Department of the Interior to Herr Landrath Grasso, July 28, 1856, SAPad A841.

65. Concerns had been and would continue to be about hygiene as well. In the 1832 cholera scare, Paderborners feared that the pile up of garbage in the promenades and along the walls threatened to spread infectious disease. SAPad A856.

66. From a letter from Königliche Polizei Inspector [Name illegible] to the Bürgermeister Herr von und zur Mühlen, November 11, 1856, SAPad A841. There is nothing in the documents to indicate if and when the gate was cleaned up.

67. Maron, "Vom Ende des Fürstbistums," 40, NWStADT M1-I-K Nr. 2297. There were many private petitions for new gates at the same time.

68. Hohmann, *Bauten des Historismus*, 101, 40. During this era, Paderborn suffered from the combined hardships of the Kulturkampf oppressions and a devastating fire, which destroyed a large swathe of the city.

69. City of Paderborn, *Bericht über die Verwaltung, . . . 1879/80.*

70. SAPad A852; GStA PK I. HA Rep. 151 Nr. 1401; City of Paderborn, *Bericht über den Stand . . . , 1884/5*, 51; SAPad A850.

71. Hohmann, *Bauten des Historismus*, 35.

72. Hohmann, *Bauten des Historismus*, 40.

73. Portions of this section were previously published in an earlier form in Poling, "Shantytowns and Pioneers."

74. Springer, *Berlin wird Weltstadt*, 148; Large, *Berlin*.

75. Hoffmann-Axthelm and Scarpa, *Berliner Mauern und Durchbrüche*, 35–36.

76. Zschocke, *Die Berliner Akzisemauer*, 75. See also Hengsbach, "Der Berliner Stadtrand"; Escher,"Stadtranderscheinungen in Berlin," 100.

77. GStA PK. I. HA Rep. 93b, Nr. 3047. The Oranienburger Gate had been moved outward in 1845–1846 and was expanded in 1849. A piece of the wall was removed in conjunction with the construction of new railway tracks in 1850. A new gate was placed between the Kottbusser and Halle Gates in 1851. GStA PK I. HA Rep 151 III, Nr. 2074 and 2075. On traffic complaints at the Hamburger and Kottbusser Gates in the 1850s and 1860s, see documents in the Landesarchiv Berlin (State Archive of Berlin; hereafter cited as LABerlin), A Pr Br Rep 0303 Nrs. 611, 622.

78. GStA PK I. HA Rep. 90a, Nr. 1800, doc.1; GStA PK I. HA Rep. 93b, Nr. 3029; GStA PK I. HA Rep. 77, Tit. 227a, Nr. 49, vol. 2; LABerlin Rep 000-02-01 Nr. 1579; GStA PK I. HA Rep. 90a, Nr. 1800. On the financial "double burden" of living in the suburbs, see also Zschocke, *Die Berliner Akzisemauer*, 40.

79. Minister of War Albrecht von Roon justified the decision with reference to 1848 in 1853. GStA PK I. HA Rep. 93b, Nr. 3029, 31.

80. Sass, *Berlin in seiner neuesten Zeit*; Dronke, *Berlin*, 13–14.

81. Bruch, *Berlin's bauliche Zukunft*, 6, 92. See also Faucher, "Die Bewegung für Wohnungsreform," 149–52.

82. Clauswitz, *Die Städteordnung von 1808*, 166–67; Hegemann, *Das steinerne Berlin*, 268; Felix Escher, *Berlin und sein Umland. Zur Genese der Berliner Stadtlandschaft bis zum Beginn des 20. Jahrhunderts* (Berlin: Colloquium Verlag, 1985), 12–13.

83. Clauswitz, *Die Städteordnung*, 167. As Otto von Gierke formulated it, the Prussian "authoritarian state" feared the "the growing big city." Quoted in Hegemann, *Das steinerne Berlin*, 268.

84. Springer, *Berlin: Die deutsche Kaiserstadt*, 64; Clauswitz, *Städteordnung*, 167–68; Springer, *Berlin: Die deutsche Kaiserstadt*, 85–86; Faucher, *Vergleichende Culturbilder*; Bruch, *Berlin's bauliche Zukunft*, 2; Faucher, "Die Bewegung für Wohnungsreform (Zweiter Artikel)," 88; Baumeister, *Stadt-Erweiterungen*, 467, 480.

85. "Berliner Schnadahupferl," *Kladderadatsch* 17, no. 58 (December 18, 1864): 231.

86. "Eine Schaar den kleinen Umstürzler," *Kreuzzeitung*, July 26, 1865, 3.

87. *Norddeutsche Allgemeine Zeitung*, July 16, 1865, 3.

88. *Norddeutsche Allgemeine Zeitung*, September 6, 1865, 3; "Illustrirte Rückblicke," *Kladderadatsch* 18, no. 44 (September 24, 1865): 177.

89. *Kreuzzeitung*, July 26, 1865, 3. On the serious business of "childish games" in the dismantlement of the wall, see also *Börsen-Zeitung*, July 23, 1865, 2691.

90. Heinrich von Trietschke (1874), quoted in Hegemann, *Das steinerne Berlin*, 409. Similarly, the illustration on the first leaf of Paul Lindenberg's book of Berlin sketches shows a young orphan with the caption "Berlin." Lindenberg, *Berlin in Wort und Bild*.

91. *Kreuzzeitung*, August 1, 1865, 2.

92. Thomason, "Prussian Police State in Berlin," 61; Davis, "'Everyday' Protest," 268; Faucher, "Die Bewegung für Wohnungsreform," 195. Alan Waterhouse provides a lively (if uneven) account, but he overstates the situation when he calls this a "border crisis." Waterhouse, *Boundaries of the City*, 262.

93. "Stadtmäuerliche Gedanke," *Kladderadatsch* 18, no. 42 (September 10, 1865): 167.

94. *Börsen-Zeitung*, September 8, 1865, 3241.

95. *Norddeutsche Allgemeine Zeitung* 4, September 6, 1865, 3.

96. *Börsen-Zeitung*, September 8, 1865, 3241; "Stadtmäuerliche Gedanken," *Kladderadatsch* 18, no. 42 (September 10, 1865): 167; also *Staatsbürger-Zeitung*, July 30, 1865, 4; *Kreuzzeitung*, September 16, 1864, 2.

97. Zschocke, *Die Berliner Akzisemauer*, 94–189; Stein, "Berlins Stadtmauer"; LABerlin A Rep 000-02-01, Nr. 1579. On the relationship between private property owners and urban development, see Forsell, *Property, Tenancy, and Urban Growth*.

98. *Norddeutsche Allgemeine Zeitung* 4, September 6, 1865, 3; *Börsen-Zeitung*, July 23, 1865, 2691; *Spenersche Zeitung*, September 3, 1865, 2.

99. GStA PK I. HA Rep. 93b, Nr. 3030.

100. Taylor, *Secular Age*, 716.

101. Lüdtke, *Police and State in Prussia*, 16.

102. Blackbourn, *History of Germany*, ch. 2.

4: THE SHANTYTOWN FRONTIER

1. Bernhardt, *Bauplatz Groß-Berlin*; Bullock and Read, *Movement for Housing Reform*. On James Hobrecht, see Strohmeyer, *James Hobrecht*.

2. Schorske, *Fin-de-Siècle Vienna*, 24; Schumacher, *Wie das Kunstwerk Hamburg*, 56; Ladd, *Urban Planning and Civic Order*.

3. Ladd, *Urban Planning and Civic Order*; Sutcliffe, *Towards the Planned City*; Fehl and Rodríguez-Lores, *Stadt-Umbau*.

4. Evans, *Death in Hamburg*; Beachy, *Gay Berlin*, ch. 2; Hett, *Death in the Tiergarten*.

5. Goff, *Shantytown, USA*, ch. 4; Vasudevan, *Metropolitan Preoccupations*, 18.

6. See, for example, Novy and Prinz, *Illustrierte Geschichte*.

7. Brian Ladd argues that this has distorted the history of urban planning and reform in Germany, a distortion he sought to redress by focusing on the cities of Cologne, Frankfurt am Main, and Düsseldorf in his own formative study. Ladd, *Urban Planning and Civic Order*, 4–5.

8. See Andrew Lees's foundational comparative study of urban thought: Lees, *Cities Perceived*, 240.

9. Lees, *Cities Perceived*, 240; Ladd, *Urban Planning and Civic Order*, 2, 13. On transatlantic conversations about urban expertise, see Rodgers, *Atlantic Crossings*; Lees, "Between Anxiety and Admiration"; Klemek, *Transatlantic Collapse*.

10. Umbach, *German Cities*.

11. On domesticity and national identity, see Reagin, "German *Brigadoon?*"; O'Donnell, "Home, Nation, Empire." On Damaschke, Bäumer, and the generation of 1890, see Repp, *Reformers, Critics*.

12. Stahr, *Nach fünf Jahren*, 12. See also Rodgers, *Atlantic Crossings*, 112–59; Umbach, *German Cities*, 1–9.

13. Bruch, *Berlin's bauliche Zukunft*, 31.

14. Stübben quoted in Ladd, *Urban Planning and Civic Order*, 114.

15. Shapiro, *Housing the Poor*; Jordan, *Transforming Paris*.

16. Sass, *Berlin in seiner neuesten Zeit*, 6; Baumeister, *Stadt-Erweiterungen*, 27; Schwabe, "Das Nomadenthum in der Berliner Bevölkerung." Hathitrust, https://babel.hathitrust.org/cgi/pt?id=mdp.39015065533914;view=1up;seq=47/.

17. Scheffler, "'Weltstadt' und 'Unterwelt.'" On cities and freedom of movement, see Zimmer, *Remaking the Rhythms of Life*, 105–8.

18. Saldern, *Häuserleben*, 40–53; Geist and Kürvers, *Das Berliner Mietshaus*.

19. Thienel, *Städtewachstum*, 84, 170.

20. Saldern, *Häuserleben*, 78–79. On the neighborhood *Kiez* or "turf" as an expression of working-class culture in Berlin, see also Swett, *Neighbors and Enemies*, 25–31.

21. Saldern, *Häuserleben*, 83, 80, 90, 81; Novy and Prinz, *Illustrierte Geschichte*, 24; Schröder, "Großstadtklang."

22. Brian, "Art from the Gutter," 36–37.

23. Ladd, *Urban Planning and Civic Order*, 83; Bernet, "Hobrecht Plan," 402.

24. Bernet, "Hobrecht Plan," 403–4.

25. Bruch, *Berlins bauliche Zukunft*; Bruch, "Zur modernen Entwickelung"; Bruch, "Der Straßenverkehr in Berlin"; Hegemann, *Das steinerne Berlin*.

26. Bernet, "Hobrecht Plan," 405–8.

27. Bruch, *Berlin's bauliche Zukunft*, 2; Bruch, "Zur modernen Entwickelung."

28. Kuhn, "Durcheinander Wohnen und Zonierung"; Saldern, *Häuserleben*.

29. Hobrecht,"Unsere Art zu wohnen," 34–35.

30. Bernhardt, "Soziale Mischung"; Bodenschatz, *Platz frei*. See also Saldern, *Häuserleben*, 46, 60, 121. Wolfgang Maderthaner makes a similar point about Vienna: that the relative sameness of the facades of suburban working-class tenements and bourgeois apartment buildings has helped contribute to the exclusion of the working classes from histories of the fin-de-siècle city. Maderthaner, "Outcast Vienna."

31. See, for example, Huber, "Ueber innere Colonisation"; Baumeister, *Stadt-Erweiterungen*, 21–32.

32. Saldern, *Häuserleben*, 60.

33. Portions of this section appeared in an earlier version in *Central European History* in 2014; see Poling, "Shantytowns and Pioneers."

34. Lange, *Berlin zur Zeit Bebels*, 30–31; Glatzer, *Berlin wird Kaiserstadt*, 80; Large, *Berlin*, 16; *Norddeutsche Allgemeine Zeitung*, July 4, 1872, 3, and August 25, 1872, 3; *Kreuzzeitung*, May 11, 1872, 2–3; *Gerichts-Zeitung*, July 13, 1872, 2.

35. See, for example, the description of "Waggon-, Höhle-, und Zillenbewohner," in the *Berliner Tageblatt*, Hauptblatt, May 8, 1872, 3.

36. "Die Wohnungsnoth und die Barackenstadt in Berlin," *Deutsche Allgemeine Zeitung*, July 14, 1872, supplement (*Beilage*).

37. Versions of the same story with these numbers ran in a number of local papers: *Spenersche Zeitung*, May 17, 1872; *Vossische Zeitung*, May 18, 1872, supplement; *Kreuzzeitung*, May 22, 1872, 3.

38. "Berliner Krawalle und Wohnungsnoth," *Deutsche Allgemeine Zeitung*, August 11, 1872, Beilage, *Staatsbürger-Zeitung*, June 1, 1872, 4. For an example of gardens and fences, see *Kreuzzeitung*, May 22, 1872, 3.

39. Nitsche, *Häuserkämpfe*, 54, 57; Bernhardt, *Bauplatz Groß-Berlin*, 45.

40. The shantytowns were described in three of the most widely distributed illustrated family journals with illustrations: Ring, "Ein Besuch in Barackia," in *Gartenlaube*; Hosang, "Berliner Wohnungsnoth," in *Über Land und Meer*; "Die Berliner Barackencolonie" in *Illustrirte Zeitung*. Another illustration of the 1872 shantytowns, a woodcut by Paul Meyersheim, appeared in the *Berliner Volkszeitung Sonntagsblatt* on July 14, 1872. This image is reprinted in Glatzer, *Berlin wird Kaiserstadt*, 81.

41. Hosang, "Berliner Wohnungsnoth," 14. Accounts ran in the *Staatsbürger-Zeitung*, *Kreuzzeitung*, *Vossische Zeitung*, *Spenersche Zeitung*, *Norddeutsche Zeitung*, *Berliner Volkszeitung*, *Demokratische Zeitung*, and *Berliner Tageblatt*. For Frankfurt, for example, see *Frankfurter Zeitung und Handelsblatt*, June 4, 1872, 1. For the transatlantic view, see the report in the *Chicago Tribune*, August 18, 1872, 8.

42. See, in turn, Saldern, *Häuserleben*, 58; Vasudevan, *Metropolitan Preoccupations*; Förster, Bielefeldt, and Reinhold, *Zur Geschichte des Deutschen Kleingartenwesens*; Stein, *Inseln im Häusermeer*.

43. Repp, "Adolf Damaschke and the Language of Popular Nationalism," ch. 2, in Repp, *Reformers, Critics*. On Heinrich Zille's youth in Berlin, see Fischer, *Heinrich Zille*, 12–17. On the memory of the 1870s shantytowns in the squatters' movement of the 1970s and 1980s, see Vasudevan, *Metropolitan Preoccupations*, 30.

44. Goff, *Shantytown, USA*, xiii, 92.

45. Shapiro, *Housing the Poor*, 45, 48; Vasudevan, *Metropolitan Preoccupations*, 4.

46. Ostwald, *Vagabunden*, 37; Schlör, *Nights in the Big City*, 160; Mattenklott and Mattenklott, *Berlin Transit*, 139–42.

47. Ostwald, *Vagabunden*; Ring, "Ein Abend im Asyl." In a later report from 1894, the city magistracy found that only one-sixth of the users of the city's homeless shelter were Berlin-born. Scheffler, "'Weltstadt' und 'Unterwelt,'" 165.

48. Quoted in Scheffler, "'Weltstadt' und 'Unterwelt,'" 168.

49. Scheffler, "'Weltstadt' und 'Unterwelt,'" 166–68. See also Walcker, *Die großstädtische Wohnungsnoth*.

50. "Die Berliner Barackencolonie," *Illustrirte Zeitung*, 435; Ring, "Ein Besuch in Barackia," *Die Gartenlaube*, 460; Dorn, "Ein Spaziergang ins Elend," 2.

51. Frederick Jackson Turner, "The Significance of the Frontier in American History (1893)," in Turner, *Rereading Frederick Jackson Turner*, 59.

52. Ring, "Ein Besuch in Barackia," 458.

53. "Die Berliner Barackencolonie," *Illustrirte Zeitung*,435; *Staatsbürger-Zeitung*, August 23, 1872 ("return to nature"). See also "Die Barackenstadt vor dem Kottbuser Thor," *Berliner Tageblatt*, May 27, 1872, 3.

54. *Kreuzzeitung*, May 22, 1872, 3. See also Ring, "Ein Besuch in Barackia," 460.

55. Here the conservative *Kreuzzeitung*'s May 22 account shares much with reformer Max Ring's rosy depiction in "Ein Besuch in Barackia" in the *Gartenlaube*, omitting details such as

the hired carpenters even when picking up stories from the pages of the *Gerichts-Zeitung*. The *Gerichts-Zeitung* (also a relatively conservative, government-aligned paper) tended to treat the shantytowns in a positive light, yet without the same evident misrepresentations and exaggerations of the *Kreuzzeitung* or *Gartenlaube* accounts.

56. Bullock and Read, *Movement for Housing Reform*, 75–77.

57. "Die Berliner Barackencolonie," 435. See also Hosang, "Berliner Wohnungsnoth," 14.

58. Thienel, *Städtewachstum*; Burkhard Hofmeister, "Die Siedlungsentwicklung Groß-Berlins," in *Siedlungsforschung* 1 (1983): 39–63.

59. "Die Berliner Barackencolonie," *Illustrirte Zeitung*, 435. See also *Deutsche Allgemeine Zeitung*, June 27, 1872, 1465; "Die Barackenstadt vor Kottbuser Thor," *Berliner Tageblatt*, May 27, 1872.

60. *Frankfurter Zeitung*, May 26, 1872, 1. Originally from the *Hamburger Correspondent*.

61. "Baracke. Organ für Neu-Berlin," *Kladderadatsch* 25, no. 33 (July 21, 1872): 130. See also Ring, "Ein Besuch in Barackia," 459; "Die Berliner Barackencolonie," 437.

62. *Deutsche Allgemeine Zeitung*, June 27, 1872, 1465.

63. Hosang, "Berliner Wohnungsnoth," 14.

64. Reprinted in the *Frankfurter Zeitung*, June 1, 1872, 1.

65. *Frankfurter Zeitung*, June 4, 1872, 1, reprinted in *Bonner Zeitung*, June 12 1872, 3.

66. "Die Berliner Barackenstadt," *Bonner Zeitung*, June 12, 1872, 3. Universität Bonn Digitale Sammlungen https://zeitpunkt.nrw/ulbbn/date/day/4789403?d=1872-06-12/.

67. Nitsche, *Häuserkämpfe*, 54.

68. *Börsen-Zeitung*, August 22, 1872, 6. See also Nitsche, *Häuserkämpfe*, 56; Damaschke, *Die Bodenreform* (1915), 462; also http://kreuzberger-chronik.de/chroniken/2012/oktober/geschichten.html/, from *Berlinische Monatsschrift*, 8, no. 97.

69. *Staatsbürger-Zeitung*, May 17, 1872; "Die Barackenstadt vor dem Kottbuser Thor," *Berliner Tageblatt*, May 27, 1872, 3. For the socialist politician and journalist Wilhelm Hasselmann, the presence of these flags convinced him that the shanty dwellers were politically a lost cause. *Vossische Zeitung*, June 11, 1872, 4th supplement, 2.

70. *Staatsbürger-Zeitung*, June 1, 1872, 4; *Demokratische Zeitung*, June 9, 1872. The Prussian flags and conservative patriotism of the squatters remained a key feature in later accounts of the shantytowns. See Julius Rodenberg, "Das Werden und Wachsen unsrer Stadt" (June 1884) in Rodenberg, *Bilder aus dem Berliner Leben*, 237–38.

71. Stein, *Inseln im Häusermeer*, 271.

72. Ward, *Autonomy, Solidarity, Possibility*, 103.

73. Urban, "Hut on the Garden Plot," 223–26.

74. Nilsen, *Working Man's Green Space*; Stein, *Inseln im Häusermeer*; Meller, "Citizens in Pursuit of Nature," 85.

75. Stein, *Inseln im Häusermeer*, 46–51, 69. See Schreber,"Die Jugendspiele."

76. Stein, *Inseln im Häusermeer*, 238, 240; Nilsen, *Working Man's Green Space*, 72, 73; Tis-

sot, *Reportagen aus Bismarcks Reich*, 242. See also *Familiengärten und andere Kleingartenbestrebungen*, 228–29.

77. Förster, Bielefeldt, and Reinhold, *Zur Geschichte des Deutschen Kleingartenwesens*, 28; Stein, *Inseln im Häusermeer*, 251–52.

78. Stein, *Inseln im Häusermeer*, 242–43.

79. Meller, "Citizens in Pursuit of Nature," 81.

80. Stein, *Inseln im Häusermeer*, 57–59, 68–70; Nilsen, *Working Man's Green Space*, 74; *Familiengärten und andere Kleingartenbestrebungen*, 228.

81. For an account of this *Generalpachtsystem*, see Nilsen, *Working Man's Green Space*, 76–77. Examples of city land temporarily leased to allotment gardeners included the Treptower area, with forty-eight hundred plots, the Rixdorf with eight hundred, and Boxhagen-Rummelburg with eight hundred. *Familiengärten und andere Kleingartenbestrebungen*, 228.

82. *Familiengärten und andere Kleingartenbestrebungen*, 229–30. See also Nilsen, *Working Man's Green Space*, 77–78.

83. Damaschke (1920) quoted in Stein, *Inseln im Häusermeer*, 89.

84. Nilsen, *Working Man's Green Space*, 74, 133–35.

85. Nilsen, *Working Man's Green Space*, 75–76; Urban, "Hut on the Garden Plot," 228, 229; Jackisch, "Nature of Berlin," 315.

86. Fallada, *Kleiner Mann—was nun?*

87. See *Der Stadtrandsiedler: Interessen-organ der Klein-Siedler des Berliner Ostens Adlershof* (1932). The first issue of this magazine defined the movement as a search for a new form of life on the edge of the city. Gohr, "Der Stadtrandsiedler."

88. Jackisch, "Nature of Berlin," 318–19.

89. Förster, Bielefeldt, and Reinhold, *Zur Geschichte des Deutschen Kleingartenwesens*, 28. See also Urban, "Hut on the Garden Plot," 243.

90. Wilson, *German Forest*, 97–98.

91. Hegemann, *Das steinerne Berlin*, 345; Erbe, "Berlin im Kaiserreich," 704–5; Bullock and Read, *Movement for Housing Reform*, 156; Bernhardt, *Bauplatz Groß-Berlin*, 40.

92. Wilson, *German Forest*, 110–24.

93. Arminius (pseudonym for Adelheid von Dohna-Poninski), *Die Großstädte*, 143, 45.

94. Rodgers, *Atlantic Crossings*, 174.

95. On excursion culture (*Ausflugskultur*) in Berlin, see Gandy, *Fabric of Space*, ch. 2, esp. 69.

96. See Gandy, *Fabric of Space*, ch. 2. There were exceptions to this rule. Helen Meller points out that a few German planners and garden theorists, such as Fritz Schumacher and Leberecht Migge, anticipated the incorporation of private gardening spaces for workers within public parks. Meller, "Citizens in Pursuit of Nature," 88–89.

97. Brian, "Art from the Gutter," 35; Ranke, *Vom Milljöh ins Milieu*, 83. On Zille's childhood and move to Berlin, see Fischer, *Heinrich Zille*, 12–16; Ranke, *Vom Milljöh ins Milieu*.

98. On Max Liebermann's connection to Zille, see Kaufhold, *Heinrich Zille*, 23.

99. Di Palma, *Wasteland*, 36–37.

100. Goff, *Shantytown USA*; McNeur, *Taming Manhattan*; Jindrich, "Shantytowns of Central Park West"; Rideing, "Squatter Life in New York." For a more detailed examination of the ways in which coverage of the 1872 shantytowns evoked tropes of the German abroad, see Poling, "Shantytowns and Pioneers." On mass migration as an imagined collective experience and conservative form of revolution, see Walker, *Germany and the Emigration*.

5: URBAN HISTORIES AND NATIONAL FUTURES IN THE GERMAN EMPIRE

1. Sombart, "Der Begriff der Stadt," 1. On the persistence of images of the medieval and Renaissance cities, see Reinborn, *Städtebau*, 12.

2. Krabbe, *Die deutsche Stadt*, 71. On the fascination with history in the German Empire, see Jefferies, *Imperial Culture*, ch. 3.

3. Sheehan, "Liberalism and the City"; Mettele, "Burgher Cities"; Palmowski, *Urban Liberalism*; 43–68; Lees, "Civic Pride of the German Middle Classes," 41–59.

4. Jenkins, *Provincial Modernity*, 2–5; Frisby, *Cityscapes of Modernity*, 130. On fairs, see Zelljadt, "Presenting and Consuming the Past."

5. Sombart, *Die deutsche Volkswirtschaft*, 8–9, 17–18. Adolf Weber makes a similar move in his 1908 book on "the big city," contrasting nightmarish accounts of the contemporary metropolis with a detailed recounting of the squalor and suffocation of the city that had preceded it. Weber, *Die Großstadt*, 1–7.

6. Friedrich Lenger, "Die Gewerbegeschichtsschreibung der Historischen Schule," in Lenger, *Sozialwissenchaft um 1900*, 31–40.

7. Koshar, *Transient Pasts*, 4.

8. Umbach, *German* Cities, 17, also 37–43.

9. Brockmann, *Nuremberg*.

10. On Nuremberg in the sixteenth century, see Strauss, *Nuremberg in the Sixteenth Century*. For a description of Nuremberg's fortifications, see, among others, Schwemmer, *Die Stadtmauer von Nürnberg*; Bruder, *Nürnberg als bayerische Garnison*; Schaefer, "Mauern und Thore."

11. To the art historian Moritz Thausing, Nuremberg looked as if "noch Mutter Natur allmächtig waltete; dieser bunter Wald von Türmen erscheint wie von selbst aus dem Boden gewachsen" (mother nature still prevailed here; this colorful forest of towers appears as if grown from the ground itself). Thausing, "Eine unblütige Commune," 70. A city history published in 1861 described Nuremberg as a settlement that grew from the earth as if "sown" (*besät*). Marx, *Geschichte der Reichstadt Nürnberg*, 5.

12. Marx, *Geschichte der Reichstadt Nürnberg*, 1; Ghillany, *Nürnberg historisch und topographisch*, 101.

13. Giesecke, "Die Befestigungen der Stadt Nürnberg"; Smith, "Nuremberg and the Topographies of Expectation." On the actual functioning of Nuremberg's walls as a border, see Newhouse, "Outside the Walls." An article in the Germanic Museum's journal lists festivals as

one reason to keep the walls. R[ockinger],"Einlegung der Stadtmauer." On Nuremberg's folk festivals, see Brockmann, *Nuremberg*, 52–55.

14. Götz, *Um Neugotik und Nürnberger Stil*, 6; Brix, "Das Problem der Vorstädte," 110; Bach, *Architektur-Skizzen aus Nürnberg*; Ritter and Dohme, *Malerische Ansichten aus Nürnberg*. In the late nineteenth century, Nuremberg's walls were also frequently photographed. See, for example, Schmidt, *Nürnberg, 1865–1909*; Hahn, *Die Stadtmauern von Nürnberg*; Kirchensittenbach, *Die Stadtmauer von Nürnberg*.

15. Kusch, *Nürnberg: Lebensbild einer Stadt*, 393–94. On Nuremberg's industrial development, see Beer, Glaser, and Winkel, *In die neue Zeit*; Bühl-Gramer, *Nürnberg 1850 bis 1892*. On the city council's efforts to remove the fortifications, see Brix, "Das Problem der Vorstädte." On the founding of the German National Museum and its character as a patriotic national endeavor first under Baron Hans von Aufseß and then Essenwein, see Brockmann, *Nuremberg*, 60–71.

16. Schwemmer, "Die Stadtmauer von Nürnberg"; Bühl-Gramer, *Nürnberg 1850 bis 1892*, 208–18. Norbert Götz examines the tension between the romanticization of Nuremberg's past and its development as a modern industrial city in the nineteenth century. Götz, *Um Neugotik und Nürnberger Stil*, 130–33, 150–52.

17. "Die Zerstörung der Nürnberger Stadtmauer," *Deutsche Bauzeitung* 5, no. 1 (January 5, 1871), 6. See also "Ein Edelstein unter Deutschlands Städten. Nürnberg," *Ueber Land und Meer* 15 (1867), 231. In the national press, Nuremberg's walls were frequently referred to as the city's *Eigentümlichkeit* (meaning "peculiarity" or "medievalism"). See, for example, "Denkmäler Frevel in Nürnberg," *Organ für christliche Kunst* 22, no. 19 (October 1, 1872): 227; "Alt-Nürnbergs Untergang," *Zeitschrift für bildende Kunst* 7 (1872): 318; Friedrich Schlutheiß, "Die Mauern um Nürnberg," *Illustrirte Zeitung* 1474 (September 30, 1871); 251.

18. Allmers called for the imperial government to intervene, taking possession of the walls as property of the Reich. Brix, "Das Problem der Vorstädte," 127; Schwemmer, "Die Stadtmauer von Nürnberg," 438–39. Allmers's allegation was originally published in the *Weserzeitung* and reprinted in the *Weimarer Zeitung* in 1875. "Die Mauerbrecher von Nürnberg," *Weserzeitung* (November 11, 1875).

19. General conservator quoted in Brix, "Das Problem der Vorstädte," 119. See also Dr. Zahn, "Ueber die Stadtmauern von Nürnberg," *Fränkische Kurier* 36, no. 134 (May 14, 1869): 1; Thausing,"Eine unblütige Commune," 74. On Thausing, see Gubser, *Time's Visible Surface*, 142.

20. H[erbert] K[önig], "Schloß Scharfenberg bei Meißen," *Gartenlaube*, no. 2 (1871): 26; Essenwein, "Review of *Architektur-Skizzen aus Nürnberg*, by Max Bach." A poem that appeared in the *Kladderadatsch* highlighted this theme, in the voice of the walls themselves chastising irresponsible residents. "An die Mauernbrecher," *Kladderadatsch* 28, no. 57 (December 12, 1875): 227. See also Lübke, "Die Nürnberger Stadtmauern." Max Schassler accused the city of hiding behind "alleged community interests." Schassler, "Kunstbarbarei," 78. The *Deutsche*

Bauzeitung similarly dismissed the housing concerns as "empty phrases." "Die Zerstörung der Nürnberger Stadtmauer," *Deutsche Bauzeitung* 5, no. 1 (January 5, 1871): 6.

21. Thausing, "Eine unblütige Commune," 69.

22. Thausing, "Eine unblütige Commune," 70.

23. "Frevel am Reliquienkästlein des deutschen Reiches," *Gartenlaube* 17 (1873), 284; August Essenwein, "Review of *Architektur-Skizzen aus Nürnberg* by Max Bach," *Anzeiger für Kunde der deutschen Vorzeit* 4 (April 1869): 116, 115. Rudy Koshar observes that in the national preservation movement after 1890, preservationists not infrequently couched their claims "in religious metaphors of piety." Koshar, "Frightful Leveler," 19.

24. In this way Nuremberg's walls operated somewhat like the Cologne Cathedral itself, as Thomas Nipperdey has described it. Thomas Nipperdey, "Der Kölner Dom als Nationaldenkmal," in Nipperdey, *Nachdenken über die deutsche Geschichte*, 156–71. Contemporaries also compared Nuremberg's walls to the Cologne cathedral. See Schultheiß,"Die Mauern um Nürnberg," 254.

25. J[so] Krsnjavi, "Die Vandalen in Regensburg und Nürnberg," *Kunst-Chronik: Beiblatt zur Zeitschrift für bildende Kunst* 15, no. 2 (1879): 18. See also Faber, *Die Zukunft Nürnbergs*. Stromer and others nominally supported anti-Catholic policies, although without enthusiasm. Lutz, *Gesellschaft und Wirtschaft*, 154. On Nuremberg's ruling elites see Schmuhl, *Die Herren der Stadt*.

26. Endres and Fleischmann, *Nürnbergs Weg in die Moderne*, 167–68.

27. Bach, "Die Mauern Nürnbergs."

28. Hirschmann, "Fortleben Reichsstädtischen Bewusstseins?" 217–33; Endres and Fleischmann, *Nürnbergs Weg in die Moderne*, 167–68. See for example Kirchensittenbach, *Die Stadtmauer von Nürnberg*. The local dialect of "Nurembergish," which continued to be spoken by the city patriciate (Hirschmann, "Fortleben Reichsstädtischen Bewusstseins," 220), became associated with the movement to preserve the walls in the publication of several works of verse supporting preservation written in the local tongue. See K. D. F., *Doppelgespräch zweier echter Nürnberger*; Friedrich Lehmann, "An die Mauernbrecher in Nürnberg," in Lehmann, *Aus dem Nürnberg Volksleben*, 183–86. The patriciate, however, was by no means uniformly in support of preservation. Nuremberg's mayor Otto von Stromer, a tireless advocate of the urban expansion projects that required removal of the walls, was himself a descendant of one of Nuremberg's most distinguished families.

29. Germanisches Nationalmuseum, Graph. Sammlung, Hist. Blätter, Kapsel 1067. The petition itself is a work of art, including a watercolor by Ritter of the moats by the Spittler Gate and written in elaborate lettering with red decorative capitals. The fifty-two signatures include those of painters, engravers, architects, academics, sculptors, photographers, metalworkers, and a handful of "art lovers" drawn from Nuremberg's business and professional classes. As far as I can tell, the signatures are all from Nuremberg residents.

30. Dr. Zahn, "Ueber die Stadtmauern," *Fränkischer Kurier* 36, no. 132 (May 12, 1869), 1.

The art historian and editor of the *Deutsche Kunst Zeitung*, Max Schassler, criticized the museum for not taking a stronger stance against the all-leveling policies of Nuremberg's Mayor Otto von Stromer. Max Schassler, "Kunstbarbarei," *Deutsche Kunst Zeitung* 17, no. 10 (March 10, 1872): 79.

31. Evident in local publications like K. D. F., *Doppelgespräch zweier echter Nürnberger*, as well as the coverage in the *Frankish Courier*.

32. Lutz, *Gesellschaft und Wirtschaft*, 112, 251; Brix, "Das Problem der Vorstädte," 124. Mayor Stromer believed most of those who suffered in overcrowded housing were themselves to blame for their fate. Endres and Fleischmann, *Nürnbergs Weg in die Moderne*, 51.

33. Representative Dr. Günther (Berlin) at the February 7, 1883 meeting of the German Reichstag in *Stenographische Berichte*, 2:1319; Brix, "Das Problem der Vorstädte," 126–27; Mittelfränkischer Architekten- und Ingenieur-Verein, *Denkschrift*; Blessing, "Der Schein der Provinzialität," 102. There followed a series of laws in the 1890s on historical preservation that effectively protected not just Nuremberg's walls but many buildings and structures in their vicinity. Bauernfeind, *Bürgermeister Georg Ritter von Schuh*, 246–47.

34. Ladd, "The Closed versus the Open Cityscape," 66.

35. Krsnjavi, "Die Vandalen," 19.

36. "Alt-Nürnbergs Untergang," *Zeitschrift für bildende Kunst* 7 (1872), 319.

37. *Beilage zum Anzeiger für Kunde der Deutschen Vorzeit* 21, no. 7 (July 1874): 230; Koshar, "Frightful Leveler." On Schwabach's defortification, see Schlüpfinger, *Schwabach*, 3–4, 172–75.

38. Baumeister, *Stadt-Erweiterungen*, 114. See also Henrici, *Von welchen Gedanken*, 5.

39. "Nivellirungs-Vandalismus in Rathhausen," *Fliegende Blätter* 54, no. 1344 (1870): 126; Meckseper and Siebenmorgen, *Die alte Stadt*, 5–7; Sutcliffe, *Towards the Planned City*, 4; Jerram, "Bureaucratic Passions."

40. Vocal advocates for the preservation of walls and gates such as Moritz Thausing and Wilhelm Lübke were significant figures in the Renaissance enthusiasm of these years. Wilhelm Lübke, whose 1872 tome on the German Renaissance was one of the principal documents of this movement, reserved significant place for city gates in his account of the great heritage of German architecture. Lübke, *Geschichte der deutschen Renaissance*, 212. See also Jefferies, *Imperial Culture*, 96.

41. Prussian Landtag, Haus der Abgeordneten, Meeting of March 12, 1894, in *Stenographische Berichte*, 1098.

42. Pelc, *Im Schutz von Mauern und Toren*, 91–92.

43. *Kölnische Volkszeitung*, November 28, 1881, 1, quoted in "Köln. Die Thorburgen," *Jahrbücher des Vereins von Alterthumsfreunden im Rheinlande* 72 (1882): 133; Koshar, *Transient Pasts*, 9.

44. Prussian Landtag, Haus der Abgeordneten, meeting of March 12, 1894, in *Stenographische Berichte*, 1094.

45. [Josef Stübben], "Das Sternthor in Bonn," *Deutsche Bauzeitung* 28, no. 42 (May 26, 1894): 259.

46. Koshar, *Transient Pasts*, 70–71.

47. Busso von der Dollen, "Der Kampf um das Sterntor: Die Auseinandersetzungen um Abriß oder Erhaltung der letzten mittelalterlichen Torburg Bonns im 19. Jahrhundert," *Bonner Geschichtsblätter* 31 (1979): 116.

48. Prussian Landtag, Haus der Abgeordneten, meeting of March 12, 1894, in *Stenographische Berichte*, 1098.

49. Faucher, *Vergleichende Culturbilder*, 1–2, 3.

50. List, *National System of Political Economy*, 101. See also Bayer, *Vom Gedanken zur Wirklichkeit*; "Aufruf an unsere Mitbürger in Sachsen die Anlage einer Eisenbahn zwischen Dresden und Leipzig betreffend" (Leipzig: W. Haack, 1834) reprinted in Bayer, *Vom Gedanken zur Wirklichkeit*, 26.

51. Ladd, *Urban Planning and Civic Order*, 186.

52. Faucher, "Die Bewegung für Wohnungsreform (Zweiter Artikel)," 88.

53. Faucher, "Die Bewegung für Wohnungsreform," 163. A number of other city planners and commentators on the housing question, such as Reinhard Baumeister and Ernst Bruch, made similar arguments about the significance of fortifications in the stunted history of the German city.

54. Faucher, *Vergleichende Culturbilder*, 1–2.

55. Schmoller, *Grundriß der Allgemeinen Volkswirtschaftslehre*; Bücher, *Entstehung der Volkswirtschaft*.

56. Bücher, *Die Entstehung der Volkswirtschaft*, 373.

57. For a useful account of the political relationship between the city and the state as it evolved over the nineteenth century, see Mettele, "Burgher Cities."

58. Jefferies, *Imperial Culture*, 97.

59. Wuttke, *Die deutschen Städte*, xii, xv; Adickes and Beutler, *Die sozialen Aufgaben*.

60. Hugo Preuß, *Staat und Stadt*, Vorträge der Gehe-Stiftung zu Dresden 1 (Leipzig: Verlag B. G. Teubner, 1909), 19, 12.

61. Goerlitz, *Die Landeshauptstadt Oldenburg*, 38–39; Hegemann, *Das steinerne Berlin*.

62. City constitutions as precedent for Weimar law is a topic explored in Sandro Mezzadra's excellent article contrasting Preuß's emphasis on the city and self-government with Max Weber's overwhelming emphasis on the territorial state as the sole source for political legitimacy. Mezzadra, "Zur Theorie der Stadt," 85–99.

63. See Bruch, "Zur modernen Entwickelung," 279.

64. Lees, *Cities Perceived*, 142–48. For the work of this school, see for example, Ludwig Bauer, *Der Zug nach der Stadt*.

65. Ratzel, "Der Lebensraum." On the Lebensraum concept, see Smith, "Friedrich Ratzel." For a discussion of Ratzel's role in the racialization of German nationalism, see Smith, *Continuities of German History*, 192. Kevin Repp and Woodruff Smith both give accounts of

Ratzel's thinking that fail to account for his (qualified) celebration of the modern big city and his understanding of the importance of big cities to national settlement. Repp, *Reformers, Critics*, 245–46; Smith, "Friedrich Ratzel."

66. Ratzel, "Die geographische Lage der großen Städte," 72. Ratzel's student Kurt Hassert elaborated similar ideas in his writings on the modern big city. Hassert, *Die Städte geographisch betrachtet*, 19.

67. Ratzel, "Die geographische Lage der großen Städte," 41–42, 37.

68. Ratzel, *Völkerkunde*, 106; Ratzel, "Die geographische Lage der großen Städte," 38, 69; Ratzel, *Städte- und Culturbilder aus Nordamerika*, 4.

69. Ratzel, *Völkerkunde*, 106.

70. For example, Countess Adelheid von Dohna-Poninski wrote in 1874: "The circumstance that, in the course of the next century, the advancing colonization efforts will offer opportunities diverse and on a grand scale to establish new cities in other parts of the world, and to restore or expand existing ones, heightens the necessity and importance of a healthy theory of the architecture of the great cities, and of the cities themselves." Arminius, *Die Großstädte in ihrer Wohnungsnoth*, 10.

71. Kollmann, *Auf deutschem Boden in Afrika*, 27–28.

72. Groeneveld, "Far Away at Home in Qingdao"; Celik, *Empire, Architecture, and the City*.

73. Kopp, *Germany's Wild East*. On the activities of the Königlich Preussische Ansiedlungskommission in den Provinzen West Preussen und Posen (Prussian Settlement Commission), founded in 1886, see especially Eddie, "Prussian Settlement Commission."

74. Ausschusse des Verbandes Norddeutscher Festungsstädte, *Erstrebung gerechter*.

75. Called the Association of German Fortified Cities after 1871. The association's membership included most of the fortified cities in the North German Federation, including Cologne, Mainz, Koblenz, Saarlouis, Spandau, Stettin (represented by James Hobrecht), Thorn, and Danzig. Stadtarchiv Mainz Signatur 70/15252, Verein deutscher Festungsstädte, folder 1.

76. GStA PK I. HA Rep. 90a No. 4571. Posadowsky-Wehner also wrote about Berlin's housing problems as a "Kulturproblem." Posadowsky-Wehner, *Weltwende*.

77. Letter to Kriegsminister Heinrich von Goßler from Richard Witting, August 16, 1900, GStA PK I. HA Rep. 90a Nr. 4571, 234–35.

78. GStA-PKB 90a I. HA 90a Nr. 4572, 17.

79. Thum, "Imperialists in Panic," 137–39.

80. Repp, *Reformers, Critics*, 69.

81. Nietzsche, "On Truth and Lies in a Nonmoral Sense," 146.

82. Ciancia, "Borderland Modernity"; also Penny, "Fashioning Local Identities."

CONCLUSION

The epigraph is from Viebig, *Die vor den Thoren*, 446.

1. On wastelanding, see Voyles, *Wastelanding*.

2. Viebig, *Die vor den Thoren*, 447.

3. Zahra, *Great Departure.* See also Zimmer, *Remaking the Rhythms of Life*, on the importance of journeys to nineteenth-century spatial organization.

4. Kleinberg, Scott, and Wilder, *Theses on Theory and History.*

5. On the influence of American social theory and preoccupations on Walker's use of the term "hometown," see Mintzker, "Paradox of Visual and Material Cultures." On the untranslatability of Walker's foundational work on German towns into German, see Friedrichs, "How German Was the German Home Town?" 489.

6. Osterhammel, *Die Verwandlung der Welt.*

7. Viebig, *Die vor den Thoren.*

8. Briel, *Innere Westvorstadt*, 11.

9. Spieske, *Erinnerungen eines alten Oldenburgers*, 1.

10. Osterhammel, *Die Verwandlung der Welt*, 465, 471, 472.

11. In Fritz Schumacher's memorably expressed trinity of urban ills: "spiritual wildness [*Verwilderung*], social wildness, aesthetic wildness." Schumacher, *Köln*, 16.

12. Slotkin, *Fatal Environment*, 33.

Bibliography

ARCHIVAL SOURCES

Geheimes Staatsarchiv Preußischer Kulturbesitz (GStA PK)—Secret State Archives of the Prussian Cultural Heritage Foundation

GStA PK I. HA Rep. 77 — Ministerium des Innern, Tit. 227a

Nr. 49, vol. 2 — Acta betreffend die Ermittelung und Feststellung des Weichbildes der Stadt Berlin, March 2, 1855–August 26, 1867

GStA PK I. HA Rep. 90a — Staatsministerium jüngere Registratur

Nr. 1799 — Akten betreffend die Erhaltung der Stadtmauern, Thore, Thürme, usw, 1830–1920

Nr. 1800 — Akten betreffend die Erhaltung der Stadtmauern, Thore, Thürme, usw, 1830–1920

Nr. 4571 — Akten betreffend die Festung Posen, vol. 1, 1853–1900

Nr. 4572 — Akten betreffend die Festung Posen, vol. 2, 1901–1913

GStA PK I. HA Rep. 93b — Ministerium der öffentlichen Arbeiten

Nr. 3029 — Akten betreffend den Bau und die Unterhaltung der Stadtmauer, Thore, Schlagbäume, Wach- und Thorhäuser in Berlin, October 7, 1852–December 10, 1859

Nr. 3030 — Akten betreffend den Bau und die Unterhaltung der Stadtmauer, Thore, Schlagbäume, Wach- und Thorhäuser in Berlin, November 2, 1865–October 24, 1866

Nr. 3047 — Acta betreffend die Beibehaltung oder Abtragung der Stadtmauern der Haupt- und Residenzstadt Berlin, July 4, 1847

GStA PK I. HA Rep. 151 — Finanzministerium IC

Nr. 1401 — Acta betreffend Stadtmauer in Paderborn, 1848–1882

GStA PK I. HA Rep. 151 — Finanzministerium III

Nr. 2050 — Acta betreffend den Durchbruch von Pforten und Oeffnungen durch die Stadtmauern, 1820–1833

Nr. 2051 — Acta betreffend den Durchbruch von Pforten, Oeffnungen durch die Stadtmauern, 1833–1877

Nr. 2158 — Acta betreffend die Erhaltung der Stadtmauern, Thore, Graben und Wälle, im Bezirk des Provinzialsteuer Directorats zu Münster, vol. 1, 1820–1836

Nr. 2159 — Acta betreffend die Erhaltung der Stadtmauern, Thore, Graben und Wälle, im Bezirk des Provinzialsteuer Directorats zu Münster, vol. 2, 1836–1881

Germanisches Nationalmuseum—German National Museum

Graph. Sammlung, Hist. Blätter

Kapsel 1067 (Hz. 2249)	Lorenz Ritter, "Blick durch den Spittlertorgraben auf die Burg," watercolor with petition, ca 1886

Landesarchiv Berlin—Das Staatsarchiv des Landes Berlin (LABerlin)—State Archive of Berlin

A Rep 000-02-01, Nr. 1579	Acten der Stadtverordneten Versammlung zu Berlin betreffend der Abbruch der hiesigen Stadtmauer, sowie im Allgemeinem die Regulierung, Pflasterung und Unterhaltung der hierdurch frei werdenden Boulevardstraßen, 1859–1882

Landesarchiv Nordrhein-Westfalen Abteilung Ostwestfalen-Lippe, Staats- und Personenstandsarchiv Detmold—North Rhine–Westphalia State Archive in Detmold (NWStADT)

M1-I-K Regierung Minden; Bauwesen

Nr. 2295	Die Stadtmauern und -Thore zu Paderborn, vol. 1, November 1816–June 1843
Nr. 2296	Die Stadtmauern und -Thore zu Paderborn, vol. 2, July 1846–December 1857
Nr. 2297	Die Stadtmauern und -Thore zu Paderborn, vol. 3, January 1858–December 1867
Nr. 2298	Die Stadtmauern und -Thore zu Paderborn, vol. 4, January 1868–1906

Niedersächsisches Landesarchiv Oldenburg (NLA OL)—Lower Saxon State Archive Oldenburg

K-ZE Best. 298 OL A Nr. 1	"Vor dem Damm-Thor zu Oldenburg im Jahre 1820"
K-ZE Best. 298 OL B Nr. 120	"Grundriss von der Stadt Oldenburg" (1774)
K-ZE Best. 298 OL C Nr. 303	"Vor dem Heil.-Geist Thor zu Oldenburg" (1830)
20-33c Nr. 6	Akten betr. die Übernahme gewißer Stücke des Hauptwalls und Grabens, seitens des Oldenburgischen Stadtmagistrats beim Verkaufe der hiesigen Festungs-werke durch das Königl. Dänische Generalkriegs Directorium und spätere Differenzen wegen der Unterhaltung der Brücken beim blauen Hause und dem Dammthor usw, 1765–1774
31-4 41 Nr. 14	Veränderungen am Heiligen-Geist und am Haaren Tor, 1792–1800

31-4 42 Nr. 10	Abtragung des Walles von der Stau-Wache bis zur Kasematte vor der Bastion am Heil.-Geist-Tor, 1800
31-13 66 Nr. 48c	Wall zwischen dem Haarenufer und der Gartenstraße
Theaterwall. Bauplätze, 1838–1857	
71-3 Nr. 63	Convolut betr. den Verkauf der Vestungswerke und andere herrschaftlichen Gebäuden in Oldenburg und Apen, und was sonst wegen des am hiesigen Magistrat abgetretenen Walls und andere Stücken vorgekommen ist, 1763–1797
71-3 Nr. 68	Acta betr. die Veränderung des Walls und der Brücke am Haaren Thor, ferner den Abfluß von der Pförtnereÿ und Anstellung eines Gehülfen des Pförtners, 1796–1802
262-1A	Historischer Altbestand des Stadtarchivs Oldenburg
262-1A Nr. 1015	Änderung der Grenzen der Stadtgemeinde Oldenburg, 1829–1854
262-1A Nr. 1016	Acta betr. die Vereinigung der Häuser außer dem Eversten und Heiligengeistthor mit der Stadt, 1816–1829
262-1A Nr. 1074	Änderung der Grenzen der Stadt Oldenburg, 1843–1874
262-1A Nr. 2185	Acta betr. die Abtragung der Staulinie, 1800–1810
262-1A Nr. 3721	Acta, betr. die Thorsperre und Aufhebung derselben, 1814–1845
262-1A Nr. 3722	Acta betreffend die Aufhebung der Thorsperre, auch Ausschreibung einer Umlage behufs Deckung des Ausfalls der Einnahme, 1844–1845
262-1Ab Nr. 6163–78	Rechnung von dem Hauptwall und Graben Königlichen Antheils so von der Stadt übernommen worden, 1765–1775
262-1P Nr. 34	Die Festungswerke der Stadt Oldenburg und deren Verkauf, 1836

Staatsarchiv Amt Leipzig (StaatAL)—Saxon State Archive Leipzig

StaatAL 20009 Nr. 4828	Acta, die Demolirung der Festungswerke beÿ der Stadt Leipzig und deren Uiberlaßung an Rath allda betrf., 1766[–1784]

Stadtarchiv Leipzig (StadtAL)—City Archive of Leipzig

Urkunden	
Urk-K. 78	Festungsbau

Ratsstube, 1097–1939	
Tit. XXIV: Gebäude	
Tit. XXIVc (K) Nr. 61	Acta, die Erneuerung und Hinauslegung des Hinterthores s.w.d.a. betrf., 1831 [1830–1845]
Tit. XXIVc (K) Nr. 65	Acta, die Anlegung eines neuen Spitalthores s.w.d.a. betrf., 1835[–1836]
Tit. XXIVc (K) Nr. 71	Acta, die äußeren Thore, Baulichkeiten daran s.w.d.a. betrf., 1823[–1861]
Tit. XXVI: Einlass in die Stadt, öffentliche und private Beleuchtung 1680–1883	
Tit. XXVI (F) Nr. 1c	Varia, den Thorschluß und die Beleuchtung betrf., 1718–1832
Tit. XXVI (F) Nr. 2c	Acta, die Abschaffung des Thorgeldes betrf., 1823[–1824]
Tit. XXVI (F) Nr. 2d	Einnahme und Ausgabebuch über das Thorgeld, 1797–1824
Tit. XXVI (F) Nr. 2e	Verschiedenes, das Ende des Thorgroschens betrf., 1824
Tit. XXVII: Anlage um die Stadt 1749–1857	
Tit. XXVII (F) Nr. 2b	Rechnungen über Austrocknung des Stadtgrabens und Unterhaltung der Garten Anlagen, 1784–1807
Tit. XXVII Nr. 7	Acta, die Anlagen um die Stadt betrf., Vols. 1–3.
Tit. XXXVIII Festungswerke 1546–1830	
Tit. XXXVIII (F) Nr. 19	Acta, die wegen Demolition derer zu der Stadt Leipzig gehörigen Festungs-Wercke, gnädigst anbefohlene Berichts-Erstattung samt was dem anhängig betrf., 1763–1785
Tit. XXXVIII (F) Nr. 29	Acta, den Stadtgraben und dessen Austrocknung betrf., 1754–1819
Tit. XXXVIII (F) Nr. 34	Acta, den Einlass in Grimmischen Thor, alhier zu Leipzig betrf., 1718–1822
Tit. XXXIX Vorstädte	
Cap. 5 Nr. 5	Acta, die vollständige Gleichstellung mit der inneren Stadt in communaler Hinsicht, 3 volumes, 1835–1848
Cap. 57b Nr. 22	Acta, das Tauchaer Thorhaus betr., 1861[–1911]

Stadtarchiv Mainz—City Archive of Mainz

70/15252: Verein deutscher Festungsstädte, Folder 1

Stadtarchiv Paderborn (SAPad)—City Archive of Paderborn

A841	Anlegung eines Tores der Rosenstraße und Verkauf von zwei Bauplätzen neben demselben, 1834–1848
A842	Pforte zwischen dem Western- und Neuhäuser Tor (Berliner Törchen), 1855–1918
A851	Stadtmauern, Tore und Türme; Pflicht zu deren Unterhaltung; Klage des Steuerfiskus c/a die Stadt Paderborn, 1831–1834
A852	Übertragung des Eigentums der Stadtmauern vom Fiskus auf die Stadt Paderborn; Klage Sander wegen des eingestürzten Steinbogens über der Pader, 1880–1920
A858	Beaufsichtigung der Promenaden, 1830–1851

PUBLISHED PRIMARY SOURCES

Periodicals

Anzeiger für Kunde der deutschen Vorzeit

Berliner Tageblatt

Berliner Volkszeitung

Börsen-Zeitung

Demokratische Zeitung

Deutsche Allgemeine Zeitung (Leipzig)

Deutsche Bauzeitung

Fliegende Blätter

Fränkischer Kurier

Gartenlaube

Gerichts-Zeitung

Germania

Illustrirte Zeitung (Leipzig)

Kladderadatsch: Humoristisch-satyrisches Wochenblatt

Kreuzzeitung

Kunst-Chronik: Beiblatt zur Zeitschrift für bildende Kunst

Leipziger Tageblatt und Anzeiger

Neue Blätter für Stadt und Land (Oldenburg)

Norddeutsche Allgemeine Zeitung

Oldenburgische Blätter

Oldenburgisches Gemeindeblatt
Organ für Christliche Kunst
Sächsische Vaterlands-Blätter
Spenersche Zeitung
Staatsbürger-Zeitung
Über Land und Meer
Vossische Zeitung
Wochenblatt herausgegeben von Mitgliedern des Architekten-Vereins zu Berlin
Zeitschrift für bildende Kunst

Books and Pamphlets

Abwehr eines Oldenburgers gegen die Verunglimpfungen seiner Vaterstadt durch Nummer 21. und 22 des diesjährigen Mindener Sonntagsblattes, das aus dem Dunklen ans Licht gezogen ist durch den Ritter vom weißen Torf. Oldenburg: Verlag von Carl Sonnenberg, 1843.

Adickes, Franz Burchard Ernst Friedrich, and Otto Beutler. *Die sozialen Aufgaben der deutschen Städte: Zwei Vorträge, gehalten auf dem ersten Deutschen Städtetage zu Dresden am 2. September 1903.* Leipzig: Verlag von Duncker and Humblot, 1903.

Arminius [pseudonym of Countess Adelheid von Dohna-Poninski]. *Die Großstädte in ihrer Wohnungsnoth und die Grundlagen einer durchgreifenden Abhilfe, mit einem Vorworte von Dr. Th. Freiherrn von der Goltz.* Leipzig: Duncker and Humblot, 1874.

Ausschusse des Verbandes Norddeutscher Festungsstädte, ed. *Erstrebung gerechter und zeitgemäßer Gesetze betreffend die Einwirkung der Festungs-Eigenschaft auf Ortschaften und Grundeigenthum. Darstellung bisher gepflogener amtlicher Verhandlungen über Festungs-Rayon-Gesetze und deren Handhabung einschließlich der Entschädigung für die daraus hervorgehenden Eigenthums-Beschränkungen.* Erfurt: Ohlenroth-schen Buchdruckerei, 1870.

Bach, Max. *Architektur-Skizzen aus Nürnberg: Malerische Radirungen.* Nuremberg: J. L. Lotzbeck's Verlag, 1871.

Bach, Max. "Die Mauern Nürnbergs. Geschichte der Befestigung der Reichsstadt." *Mitteilungen des Vereins für Geschichte der Stadt Nürnberg* 5 (1883): 47–96.

Bauer, Ludwig. *Der Zug nach der Stadt und die Stadterweiterung. Eine Rassenhygienische Studie.* Stuttgart: Druck und Verlag von W. Kohlhammer, 1904.

Baumeister, R[einhard]. *Moderne Stadterweiterungen: Vortrag.* Zeit- und Streit-Fragen: Flugschriften zur Kenntnis der Gegenwart 7. Hamburg: Verlag von J. F. Richter, 1887.

Baumeister, Reinhard. *Stadt-Erweiterungen in technischer, baupolizeilicher und wirtschaftlicher Beziehung.* Berlin: Verlag von Ernst and Korn, 1876.

"Die Berliner Barackencolonie." *Illustrirte Zeitung* 58, no. 1511 (1872): 435–37.

Beta, Hans [Heinrich Bettzeich]. *Wohl- und Uebelthäter in unseren Großstädten.* Deutsche Zeit- und Streitfragen: Flugschriften zur Kenntniß der Gegenwart 61. Berlin: C. G. Lüderitz'sche Verlagsbuchhandlung Carl Habel, 1875.

Blum, Hans. *Robert Blum: Ein Zeit- und Charakterbild für das deutsche Volk*. Leipzig: Verlag von Ernst Keil, 1878.

Bruch, Ernst. *Berlin's bauliche Zukunft und der Bebauungsplan*. Berlin: Kommissions-Verlag von Carl Beelitz, 1870.

Bruch, Ernst. "Der Straßenverkehr in Berlin." In *Berlin und seine Entwicklung*. Gemeinde-Kalender und städtisches Jahrbuch 2, edited by Heinrich Schwabe, 65–121. Berlin: Verlag von J. Guttentag, 1868.

Bruch, Ernst. "Zur Modernen Entwickelung der deutschen Hauptstadt." In *Deutsche Warte: Umschau über das Leben und Schaffen der Gegenwart* 1, edited by Bruno Meyer, 277–86. Hildburghausen: Verlag des Bibliographischen Instituts, 1871.

Bücher, Karl. *Die Entstehung der Volkswirtschaft*. Tübingen: H. Laupp'sche Buchhandlung, 1904.

Bucholtz, F[ranz]. *Aus dem Oldenburger Lande. Bilder und Skizzen*. Oldenburg: Druck und Verlag von Gerhard Stalling, 1889.

City Council of Leipzig. "Instruktion des Leipziger Rats für die Torschreiber, an den Tagen der Almosenausteilung die fremden Bettler strict abzuweisen. 26. März 1745." In *Der Leipziger Rat und die Bettler: Quellen und Analysen zu Bettlern und Bettelwesen in der Messestadt bis ins 18. Jahrhundert*, by Helmut Bräuer, 153–54. Leipzig: Leipziger Universitätsverlag GmbH, 1997.

Clauswitz, [Paul]. *Die Städteordnung von 1808 und die Stadt Berlin; Festschrift zur hundertjährigen Gedenkfeier der Einführung der Städteordnung*. Berlin: Verlag von Julius Springer 1908.

Damaschke, Adolf. *Die Bodenreform: Gründsätzliches und Geschichtliches zur Erkenntnis und Überwindung der sozialen Not*. 10th ed. Jena: Fischer Verlag, 1915.

Darstellung des von der Stadtgemeinde Oldenburg vorgelegten Planes über die Erweiterung ihrer Gränzen. Oldenburg: Druck von Gerhard Stalling, 1854.

Dieterici, Carl. *Zur Geschichte der Steuer-reform in Preußen von 1810 bis 1820: Archiv-Studien*. Berlin: Druck und Verlag von Georg Reimer, 1875.

Dorn, Friedrich. "Ein Spaziergang ins Elend." *Germania*, September 7, 1872, 2.

Dronke, Ernst. *Berlin*. First published 1846. Edited by Rainer Nitsche. Darmstadt: Hermann Luchterhand Verlag, 1974.

Duntze, Johann Hermann. *Geschichte der freien Stadt Bremen*. Vol. 4. Bremen: Johann Georg Heyse, 1851.

Essenwein, August. "Review of *Architektur-Skizzen aus Nürnberg*, by Max Bach." *Anzeiger für Kunde der deutschen Vorzeit* 4 (April 1869): 114–15.

F., K. D. *Doppelgespräch zweier echter Nürnberger: Die Einlegung ihrer alten ehrwürdigen Stadtmauer betreffend*. Nuremberg: Hessel's Druck, [1888].

Faber, Lothar von. *Die Zukunft Nürnbergs, seinen Mitbürgern gewidmet*. Nuremberg: U. E. Sebald, 1879.

Fallada, Hans. *Kleiner Mann—was nun?* 7th ed. Berlin: Aufbau Taschenbuch, 2014.

Familiengärten und andere Kleingartenbestrebungen in ihrer Bedeutung für Stadt und Land. Schriften der Zentralstelle für Volkswohlfahrt 8. Berlin: Carl Heymanns Verlag, 1913.

Faucher, Julius. "Die Bewegung für Wohnungsreform." *Vierteljahrschrift für Volkswirthschaft und Culturgeschichte* 3, no. 4 (1865): 127–99.

Faucher, Julius. "Die Bewegung für Wohnungsreform (Zweiter Artikel)." *Vierteljahrschrift für Volkswirthschaft und Culturgeschichte* 4, no. 3 (1866): 86–151.

Faucher, Julius. *Vergleichende Culturbilder aus den vier Europäischen Millionenstädten (Berlin—Wien—Paris—London).* Hannover: Carl Rümpler, 1877.

"Feier des sechsten Januar 1845 [500 Jahre Stadt Oldenburg]: [nebst] Fest-Programm." Oldenburg, ca. 1845.

Freytag, Gustav. *Die verlorene Handschrift.* First published 1864. Reprint of the 1935 edition. Karben: Verlag Petra Wald, 1997.

Fritsch, Thomas von. "Über die ersten nach dem Friedensschlüsse einzuleitende Schritte. Denkschrift Fritschs gerichtet an Premierminister Brühl, November 1761." In *Die Staatsreform in Kursachsen, 1762–1763: Quellen zum Kursächsischen Rétablissment nach dem Siebenjährigen Kriege*, edited by Horst Schlechte, 178–89. Berlin: Rütten and Loening, 1958.

"Die Gebrüder Fatalis in Schöppenstadt an den Herrn Merkur in Dresden. Den 8sten. August 1823." *Merkur. Mittheilungen aus Vorräthen der Heimath und der Fremde, für Wissenschaft, Kunst und Leben* 19 (August 18, 1823): 395–96.

German Reichstag. *Stenographische Berichte über die Verhandlungen des Reichstages, V. Legislaturperiode, II. Session.* Volume 2. Berlin: Verlag der Buchdruckerei der "Norddeutschen allgemeinen Zeitung," 1883. https:// reichstagsprotokolle.de/Blatt3_k5_bsb00018439_00001.html/.

Ghillany, F[riedrich] W[ilhelm]. *Nürnberg historisch und topographisch nach den ältesten vorhandenen Quellen und Urkunden.* Munich: Verlag von Georg Franz, 1863.

Goerlitz, Theodor, and the City of Oldenburg. *Die Landeshauptstadt Oldenburg.* Berlin-Halensee: Deutscher Architektur- und Industrie-Verlag, 1927.

Goethe, Johann Wolfgang von. *Faust I.* In *Werke*, edited by Friedmar Apel et al. Vol. 3. Frankfurt: Insel Verlag, 1998.

Goethe, Johann Wolfgang von. *Goethe's Autobiography: Poetry and Truth from My Own Life.* Translated by R. O. Moon. Washington, DC: Public Affairs Press, 1949.

Gohr, Arnold. "Der Stadtrandsiedler." *Der Stadtrandsiedler* 1, no. 1 (January 1932): 1–2.

Gretschel, C[arl] C[hristian] C[arus]. *Leipzig und seine Umgebungen.* 2nd ed. Leipzig: Friedrich Fleischer, 1836. Reprint, Zentralantiquariat der Deutschen Demokratischen Republik Leipzig 1980.

Große, Karl. *Geschichte der Stadt Leipzig von der ältesten bis auf die neueste Zeit.* 2 vols. Leipzig: Zangenberg and Himly, 1897–1898.

Gruner, Justus. *Meine Wallfahrt zur Ruhe und Hoffnung oder Schilderung des sittlichen und bür-*

gerlichen Zustandes Westphalens am Ende des achtzehnten Jahrhunderts. Frankfurt am Main: Philipp Heinrich Guilhauman, 1802.

"Gutachten der Osternburger Bevollmächtigten über die von der Stadt Oldenburg vorgeschlagene Erweiterung der Stadtgränzen." March 30, 1854. In Oldenburg County, *Oldenburger Actenstücke*, 6–21.

Hahn, Johannes. *Die Stadtmauern von Nürnberg*. Nuremberg: Heinrich Schrag, 1871.

Hassert, Kurt. *Die Städte geographisch betrachtet: Aus Natur und Geisteswelt: Sammlung wissenschaftlich-gemeinverständlicher Darstellungen*. Leipzig: Druck und Verlag von B. G. Teubner, 1907.

Hegemann, Werner. *Das steinerne Berlin: Geschichte der größten Mietskasernen der Welt*. Berlin: Verlag von Gustav Kiepenheuer, 1930.

[Heidecke, Benjamin]. *Tableau von Leipzig im Jahr 1783. Eine Skizze*. Leipzig: n.p., 1784.

Henrici, Karl. *Von welchen Gedanken sollen wir uns beim Ausbau unserer deutschen Städte leiten lassen? Vortrag gehalten im Kunst- und Gewerbeverein in Trier am 10. Januar 1894*. Trier: Verlag der Fr. Litz'schen Buchhandlung, 1894.

Hirschfeld, Christian C. L. *Theory of Garden Art*. Translated by Linda B. Parshall. Philadelphia: University of Pennsylvania Press, 2001. Originally published in five volumes as *Theorie der Gartenkunst*. Leipzig: M. G. Weidmanns Erben und Reich, 1779–1785.

Hobrecht, James. "Unsere Art zu wohnen." *Über öffentliche Gesundheitspflege* (Stettin, 1868). In *Häuserkämpfe, 1872, 1920, 1945, 1982*, edited by Rainer Nitsche, 34–35. Berlin: Transit, 1981.

Hocquél, Wolfgang. *Leipzig Architektur von der Romantik bis zur Gegenwart*. Leipzig: Passage-Verlag, 2001.

Huber, Victor Aimée. "Ueber Innere Colonisation." *Janus. Jahrbücher deutscher Gesinnung, Bildung und That* 1, no. 7 (1846): 193–222, no. 8 (1846): 225–55.

Jaspers, Karl. "Oldenburger Welt vor neunzehnhundert (1893–1899)." In *Oldenburg: Eine feine Stadt am Wasser Hunte*, edited by Hermann Lübbing, 254–60. Oldenburg: Heinz Holzberg Verlag, 1979.

Kirchensittenbach, Guido von Volckamer auf. *Die Stadtmauer von Nürnberg mit ihren Veränderungen während dreier Jahrhunderte dargestellt durch Abbildungen aus dem 17., 18. und 19. Jahrhundert*. Munich: Drück und Verlag von Dieglmayer and Fuhrmann, 1897.

Kollmann, Paul. *Auf deutschem Boden in Afrika: Ernste und heitere Erlebnisse*. Berlin: Alfred Schall, 1900.

"Köln. Die Thorburgen." *Jahrbücher des Vereins von Alterthumsfreunden im Rheinlande* 72 (1882): 132–35.

Kraus, A., and C. Weiss. *Das Stadt-Thor in Bild und Wort*. Illustrated by A. Gaber. Glogau: Verlag von Carl Flemming, 1860.

Krögen, Karl Heinrich. *Freie Bemerkungen über Berlin, Leipzig, Prag*. First published 1785. Leipzig: Gustav Kiepenheuer Verlag, 1986.

Krsnjavi, J[so]. "Die Vandalen." *Kunst-Chronik: Beiblatt zur Zeitschrift für bildende Kunst* 15, no. 2 (1879): 17–20.

Lange, Helene. *Lebenserinnerungen*. Berlin: F. A. Herbig, GmbH, 1921.

L[angermann], J[ohann] G[ottfried]. *Bemerkungen über Leipzig und einige verkannte, oder nicht genug erkannte Vorzüge und Verschönerungen dieser Stadt in Briefen von J. G. L.* Leipzig: J. G. Beygang, 1794.

Lasius, Ernst Friedrich Otto. *Blicke in der Stadt Oldenburg, Vergangenheit und Zukunft*. Oldenburg: Schnellpressendruck und Verlag der Schulze'schen Buchhandlung, 1853.

Lasius, Ernst Friedrich Otto. *Oldenburg zur Zeit unserer Väter. Eine Festgabe zur Feier des fünfhundertjährigen Bestehens unserer Stadt*. Oldenburg: Druck und Verlag von Gerhard Stalling, 1845.

Lehmann, Friedrich. *Aus dem Nürnberger Volksleben*. Nuremberg: T. Hässlein, 1882.

Leipzig: Ein Handbuch für Reisende die ihren Aufenthalt daselbst sich angenehm und nützlich machen wollen. Leipzig: Voß und Leo, 1792.

Der Leipziger Promenadenring zur Biedermeierzeit. Reprint of the 1850 edition with notes by Ursula Walter. Leipzig: Kunstverlag H. C. Schmiedicke, 1989.

Leonhardi, Friedrich Gottlob. *Geschichte und Beschreibung der Kreis- und Handelsstadt Leipzig nebst der umliegenden Gegend*. Leipzig: Johann Gottlob Beygang, 1799.

Lindenberg, Paul. *Berlin in Wort und Bild*. Berlin: Ferdinand Dümmlers Verlagsbuchhandlung, 1895.

List, Friedrich. *The National System of Political Economy*. Translated by Sampson S. Lloyd. London: Longmans, Green, 1916.

Loeffler, Emil von. *Geschichte der Festung Ulm*. 2nd ed. Ulm: Wagner, 1883.

Lübke, Wilhelm. *Geschichte der deutschen Renaissance*. Stuttgart: Verlag von Ebner and Seubert, 1872.

Lübke, W[ilhelm]. "Die Nürnberger Stadtmauern. Ein Schmerzenschrei." *Beilage zur Allgemeine Zeitung* 238 (August 26, 1871): 4211–12.

Maron, H. *Die Mahl- und Schlachtsteuer*. Berlin: Im Selbstverlage des Verfassers, Marienstr 30, 1863.

Marx, Leonhard Wilhelm. *Geschichte der Reichstadt Nürnberg*. 2nd ed. Nuremberg: J. Heerdegen (Barbeck) Antiquar, 1872. First published in 1861.

Mittelfränkischer Architekten- und Ingenieur-Verein. *Denkschrift des Mittelfränkischen Architekten- und Ingenieur-Vereins, die Einfüllung des Stadtgrabens und Einlegung der Stadtmauern Nürnbergs betreffend*. Nuremberg: W. Tümmels Buchdruckerei, 1888.

Neue Beschreibung von Leipzig: Ein Handbuch für Fremde und Einheimische, welche die Merkwürdigkeiten und Umgebungen dieser Handelsstadt näher kennen und ihren Aufenthalt zweckmähsig benutzen wollen. Leipzig: J. E. Hinrichs, 1806.

Nietzsche, Friedrich. "On Truth and Lies in a Nonmoral Sense." In *The Birth of Tragedy and Other Writings*, edited by Raymond Geuss and Ronald Speirs, translated by Ronald Speirs. Cambridge: Cambridge University Press, 1999.

Ohrt, Heinrich. *Die Großherzoglichen Gärten und Parkanlagen zu Oldenburg dargestellt in Wort und Bild*. Oldenburg: Schulzesche Hof-Buchhandlung und Hof-Buchdruckerei, 1890.

Oldenburg County. *Oldenburger Actenstücke, betreffend den von der Stadtgemeinde Oldenburg vorgelegten Plan über die Erweiterung ihrer Gränzen*. Oldenburg: Druck von Gerhard Stalling, 1854.

Oldenburgischer Staats-Kalender auf das Jahre Christi 1824. Oldenburg: Schulze, 1824.

Ostwald, Hans. *Vagabunden. Ein autobiographischer Roman*. Edited by Klaus Bergmann, Schriftenreihe des Instituts für sozialhistorische Forschung. Frankfurt am Main: Campus Verlag, 1980.

Paderborn, City of. *Bericht über die Verwaltung und den Stand der Gemeinde-Angelegenheiten der Stadt Paderborn für das Geschäftsjahr 1879/80*. Paderborn: Schöning'sche Buchdruckerei, 1880.

Paderborn, City of. *Bericht über die Verwaltung und den Stand der Gemeinde-Angelegenheiten der Stadt Paderborn für das Geschäftsjahr 1880/81*. Paderborn: Schöning'sche Buchdruckerei, 1881.

Paderborn, City of. *Bericht über die Verwaltung und den Stand der Gemeinde-Angelegenheiten der Stadt Paderborn für das Geschäftsjahr 1884/85*. Paderborn: Schöning'sche Buchdruckerei, 1885.

Philippi, Johann Jacob Martin. *Sammlung sämmtlicher neuer Preußischer Gesetze über die indirecten Steuern*. Cologne: Johann Peter Bachem, 1830.

Poppe, Franz. *Zwischen Ems und Weser: Land und Leute in Oldenburg und Ostfriesland*. Oldenburg: Schulzesche Hof-Buchhandlung und Hof-Buchdruckerei, 1888.

Posadowsky-Wehner, Arthur Adolf. *Weltwende: Gesammelte politische Aufsätze von Graf Posadowsky*. 3rd ed. Stuttgart: Walter Hädecke Verlag, 1920.

Prasch, Detlev. *Vertraute Briefe über den politischen und moralischen Zustand von Leipzig*. London: Dodsley, 1787.

Preuß, Hugo. *Staat und Stadt*. Vorträge der Gehe-Stiftung zu Dresden 1. Leipzig: Verlag B. G. Teubner, 1909.

Prince-Smith, John. "Excerpts from His Collected Writings." Translated by Jeremiah Riemer. In "From *Vormärz* to Prussian Dominance (1815–1866)," edited by Jonathan Sperber, http://germanhistorydocs.ghi-dc.org/sub_document.cfm?document_id=468/.

Prussian Landtag. *Stenographische Berichte über die Verhandlungen der durch die Allerhöchste Verordnung vom 26. Dezember 1893 einberufenen beiden Häuser des Landtages*. Vol. 2. Berlin: W. Moeser Hofbuchdruckerei, 1894.

Ratzel, Friedrich. "Die geographische Lage der großen Städte." *Die Großstädte: Vorträge und Aufsätze zur Städteausstellung, Jahrbuch der Gehe-Stiftung zu Dresden* 9 (Winter 1902/1903): 33–72.

Ratzel, Friedrich. "Der Lebensraum: Ein biogeographisches Studie." In *Festgaben für Albert Schäffle zur 70. Wiederkehr seines Geburtstages am 24. Februar 1901*, edited by Karl Bücher et al., 104–89. Tübingen: Verlag der H. Laupp'schen Buchhandlung, 1901.

Ratzel, Friedrich. *Städte- und Culturbilder aus Nordamerika*. Leipzig: F. A. Brockhaus, 1876.

Ratzel, Friedrich. *Völkerkunde: Erster Band*. 2nd ed. Leipzig: Bibliographisches Institut, 1894.

Reinick. "Resultate der Mahl- und Schlachtsteuer in der Periode von 1838 bis mit 1861." *Zeitschrift des Königlichen preussischen Statistischen Bureaus* 3, no. 9 (September 1863): 217–34.

Rideing, William H. "Squatter Life in New York." *Harper's New Monthly Magazine* (September 1880): 562–69.

Riehl, Wilhelm Heinrich. *The Natural History of the German People*. Edited and translated by David J. Diephouse. Studies in German Thought and History 13. Lewiston, NY: E. Mellen Press, 1990.

Ring, Max. "Ein Abend im Asyl für Obdachlose 1870." *Gartenlaube* 70, no. 4 (1870): 54–56.

Ring, Max. "Ein Besuch in Barackia. Berliner Lebensbild." *Gartenlaube* 72, no. 28 (1872): 458–61.

Ritter, Lorenz, and Robert Dohme. *Malerische Ansichten aus Nürnberg*. Berlin: Wasmuth, 1876.

R[ockinger], Dr. "Einlegung der Stadtmauer in Nürnberg." *Anzeiger für Kunde der deutschen Vorzeit* 22, no. 7 (July 1875): 229–30.

Rodenberg, Julius. *Bilder aus dem Berliner Leben*. Berlin: Verlag von Gebrüder Paetel, 1885.

Rose, L. W. *Der Bremer Wall*. 1860. 2nd ed. Bremen: Verlag von Hermann Gesenius, 1865.

Roth, Joseph. *Flight without End*. Translated by David LeVay. Woodstock, NY: Overlook Press, 2003.

Sass, Friedrich. *Berlin in seiner neuesten Zeit und Entwicklung*. Leipzig: Julius Koffka, 1846.

Schaefer, K. "Mauern und Thore des alten Nürnberg." *Die Baukunst* 2, no. 8 (1902): 1–17.

Schassler, Max. "Kunstbarbarei." *Deutsche Kunst Zeitung* 17, no. 1 (March 10, 1872): 78.

Scheffler, Karl. *Berlin: Ein Stadtschicksal*. Berlin-Westend: Erich Reiss Verlag, 1910.

Schemionek, F. *Die Mahl- und Schlachtsteuer in Berlin*. Berlin: Im Selbstverlage des Verfassers, 1863.

Schilling, Otto. *Innere Stadt-Erweiterung*. Berlin: Der Zirkel, 1921.

Schlechte, Horst, ed. *Die Staatsreform in Kursachsen, 1762–1763: Quellen zum Kursächsischen Rétablissment nach dem Siebenjährigen Kriege*. Berlin: Rütten and Loening, 1958.

Schmedes, L. *Das Octroi in Oldenburg aus staatsrechtlichen und nationalökonomischen Gesichtspunkten beurtheilt; nebst einer Beleuchtung der daselbst bestehenden Armentaxe. Ein Vortrag gehalten im Vereine zur Beförderung der Volksbildung in Oldenburg, am 26. April 1846*. Oldenburg: Druck und Verlag der Schulzeschen Buchhandlung, 1846.

Schmidt, Ferdinand. *Nürnberg, 1865–1909: Photographien von Ferdinand Schmidt*. Munich: Schirmer-Mosel, 1987.

Schmoller, Gustav. *Grundriß der allgemeinen Volkswirtschaftslehre*. Vol. 1, 3rd ed. Leipzig: Verlag von Duncker and Humblot, 1908.

Schreber, Daniel Gottlob Mortiz. *Ärztliche Zimmer-Gymnastik oder System der ohne Gerät und*

Beistand überall ausführbaren heilgymnastischen Freiübungen als Mittel der Gesundheit und Lebenstüchtigkeit für beide Geschlechter, jedes Alter und alle Gebrauchszwecke. 8th ed. Leipzig: Friedrich Fleischer, 1862.

Schreber, Daniel Gottlob Mortiz. "Die Jugendspiele in ihrer gesundheitlichen und pädagogischen Bedeuteung." *Die Gartenlaube* 60, no. 26 (1860): 414–16.

Schultheiß, Friedrich. "Die Mauern um Nürnberg." *Illustrirte Zeitung* 1474 (September 30, 1871): 254.

Schumacher, Fritz. *Köln: Entwicklungsfragen einer Großstadt.* With Wilhelm Arntz. Cologne: Saaleck-Verlag, 1923.

Schumacher, Fritz. *Wie das Kunstwerk Hamburg nach dem großen Brande entstand.* Veröffentlichungen des Vereins für Hamburgische Geschichte 2. Berlin: Verlag von Karl Curtius, 1920.

Schwabe, H[ermann]. "Das Nomadenthum in der Berliner Bevölkerung." *Berliner Stadtisches Jahrbuch für Volkswirthschaft und Statistik* 1 (1874): 26–37. https://babel.hathitrust.org/cgi/pt?id=mdp.39015065533914;view=1up;seq=47/.

Semmig, Hermann. "Das wilde Viertel in Leipzig." *Der Leuchtturm. Monatsschrift zur Unterhaltung und Belehrung für das deutsche Volk* 1 (1847): 17–21.

Simmel, Georg. "Die Großstädte und das Geistesleben." In *Die Großstädte: Vorträge und Aufsätze zur Städteausstellung, Jahrbuch der Gehe-Stiftung zu Dresden* 9 (Winter 1902/1903): 185–206.

Simmel, Georg. *On Individuality and Social Forms: Selected Writings.* Edited by Donald Levine. Chicago: University of Chicago Press, 1971.

Sombart, Werner. "Der Begriff der Stadt und das Wesen der Städtebildung." *Archiv für Sozialwissenschaften und Sozialpolitik* 25, no. 1 (1907): 1–9.

Sombart, Werner. *Die deutsche Volkswirtschaft im neunzehnten Jahrhundert.* Berlin: Georg Bondi, 1903.

Spieske, H. K. *Erinnerungen eines alten Oldenburgers.* Oldenburg: Druck und Verlag der Schulzeschen Buchhandlung, 1874.

Springer, Robert. *Berlin: Die deutsche Kaiserstadt nebst Potsdam und Charlottenburg mit ihren schönsten und hervorragendsten Monumenten.* Darmstadt: Verlag von Friedrich Lange, 1878.

Springer, Robert. *Berlin wird Weltstadt: Ernste und heitere Culturbilder.* Berlin: Verlag der Hausfreund-Expedition, 1866.

Stahr, Adolf. *Nach fünf Jahren. Pariser Studien aus dem Jahre 1855.* Oldenburg: Schnellpressendruck und Verlag der Schulzeschen Buchhandlung, 1857.

Starklof, L[udwig]. *Moor-Kanäle und Moor-Colonien zwischen Hunte und Ems.* Oldenburg: Druck und Verlag der Schulzeschen Buchhandlung, 1847.

Strackerjan, Ludwig. *Von Land und Leuten. Bilder und Geschichten aus dem Herzogtum Oldenburg.* Oldenburg: Schulzesche Hof-Buchhandlung und Hof-Buchdrückerei, 1881.

Thausing, Moritz. "Eine unblütige Commune." *Zeitschrift für bildende Kunst* 7 (1872): 69–74.

Tissot, Victor. *Reportagen aus Bismarcks Reich: Berichte eines reisenden Franzosen, 1874–1876.* Edited by Erich Poll. Stuttgart, 1989.

Tissot, Wilhelm. *Die Geschichte der Stadt Leipzig.* Leipzig: n.p., 1778.

Tönnies, Ferdinand. *Community and Civil Society.* Translated by Charles P. Loomis. New Brunswick, NJ: Transaction Books, 1988.

Turner, Frederick Jackson. *Rereading Frederick Jackson Turner: "The Significance of the Frontier in American History" and Other Essays with Commentary by John Mack Faragher.* New York: Henry Holt, 1994.

Viebig, Clara. *Die vor den Thoren.* Stuttgart: Deutsche Verlag, 1910.

Villaume, A. *Handbuch der preußischen Steuer- und Zoll-Gesetzgebung mit besonderer Rücksicht auf den Gebrauch bei den Gerichten.* Berlin: Verlag von G. Reimer, 1844.

Walcker, Karl. *Die großstädtische Wohnungsnoth, ihre Ursachen und Heilmittel.* Deutsche Zeit- und Streitfragen 7. Hamburg: Verlagsanstalt und Drückerei Actien Gesellschaft, 1892.

Weber, Adolf. *Die Großstadt und ihre sozialen Probleme.* Leipzig: Verlag von Quelle and Meyer, 1908.

Weidinger, Carl. *Leipzig: Ein Führer durch die Stadt und ihre Umgebungen.* Leipzig: Verlagsbuchhandlung von J. J. Weber, 1860.

Weydemeyer, Joseph. "Ueber Steuern in Preußen. Die jetzt in Preußen bestehenden Steuern." *Das westphälische Dampfboot. Eine Monatsschrift* 3. 1847. Reprint, Glashütten im Taunus: Verlag Auvermann KG, 1972.

Wustmann, Gustav, ed. *Quellen zur Geschichte Leipzigs.* Veröffentlichungen aus dem Archiv und der Bibliothek der Stadt Leipzig 1. Leipzig: Duncker and Humblot, 1889.

Wuttke, Robert, ed. *Die deutschen Städte; geschildert nach den Ergebnissen der ersten deutschen Städteausstellung zu Dresden, 1903.* Leipzig: Verlag von Friedrich Brandstetter, 1904.

SECONDARY SOURCES

Aaslestad, Katherine. "Cities and War: Modern Military Urbanism in Hamburg and Leipzig during the Napoleonic Era." *German History* 35, no. 3 (2017): 381–402.

Aaslestad, Katherine. *Place and Politics: Local Identity, Civic Culture, and German Nationalism in North Germany during the Revolutionary Era.* Studies in Central European History 36. Leiden: Brill, 2005.

Applegate, Celia. "The Mediated Nation: Regions, Readers, and the German Past." In *Saxony in German History: Culture, Society, and Politics, 1830–1933*, edited by James Retallack, 33–50. Ann Arbor: University of Michigan Press, 2000.

Applegate, Celia. *A Nation of Provincials: The German Idea of Heimat.* Berkeley: University of California Press, 1990.

Applegate, Celia. "Senses of Place." In *The Oxford Handbook of Modern German History*, edited by Helmut Walser Smith, 49–70. Oxford: Oxford University Press, 2011.

Bassin, Mark. "Turner, Solov'ev, and the 'Frontier Hypothesis': The Nationalist Signification of Open Spaces." *The Journal of Modern History* 65, no. 3 (September 1993): 473–511.

Bauer, Markus, and Thomas Rahn, eds. *Die Grenze: Begriff und Inszenierung*. Berlin: Akademie Verlag GmbH, 1997.

Bauernfeind, Martina. *Bürgermeister Georg Ritter von Schuh: Stadtentwicklung in Erlangen und Nürnberg im Zeichen der Hochindustrialisierung 1878–1913*. Edited by Werner K. Blessing, Michael Diefenbacher and Rudolf Endres: Nürnberger Werkstücke zur Stadt- und Landesgeschichte 60. Nuremberg: Stadtarchiv Nürnbergs, 2000.

Bayer, Rolf, ed. *Vom Gedanken zur Wirklichkeit: Dokumente zur ersten deutschen Ferneisenbahn von Leipzig nach Dresden, 1833–1839*. Berlin: transpress VEB Verlag für Verkehrswesen, 1989.

Beachy, Robert. *Gay Berlin: Birthplace of a Modern Identity*. New York: Alfred A. Knopf, 2014.

Beachy, Robert. "Reforming Interregional Commerce: The Leipzig Trade Fairs and Saxony's Recovery from the Thirty Years' War." *Central European History* 32, no. 4 (1999): 431–52.

Beachy, Robert. *The Soul of Commerce: Credit, Property, and Politics in Leipzig, 1750–1840*. Boston: Brill, 2005.

Beck, Hermann. *The Origins of the Authoritarian Welfare State in Prussia*. Ann Arbor: University of Michigan Press, 1995.

Beer, Helmut. "Aufbruch zur Großstadt und Arbeiterhochburg. Zur Entwickelung Nürnbergs 1850–1900." In Beer, Glaser, and Winkel, *In die neue Zeit*, 95–170.

Beer, Helmut, Hermann Glaser, and Udo Winkel, eds. *In die neue Zeit. Nürnberg 1850–1900. Eine Ausstellung des Stadtarchivs Nürnberg zum 125-jährigen Bestehen der Nürnberger Sozialdemokratie 13. Dezember 1991–29. Februar 1992*. Nuremberg: Selbstverlag der Stadt Nürnberg, 1991.

Benevolo, Leonardo. *The European City*. Translated by Carl Ipsen. Oxford: Blackwell, 1993.

Berg, Eugenie, and Detlef Hoffmann. "Das Duis-Album—eine Quelle der Moorromantik: Heimatbilder aus Moor und Heide (1911–1916)—gesammelt von dem Lehrer Johann Duis. In *Moor—eine verlorene Landschaft: Vorträge anlässlich des Symposiums in Oldenburg vom 5.–7. November 1999*, edited by Carsten Ritzau, Schriftenreihe des Landesmuseum für Natur und Mensch 20, 135–53. Oldenburg: Isensee Verlag, 2001.

Berg, Eugenie, Lioba Meyer, and Ulf Steitz. *Moderne Zeiten: Industrie- und Arbeiterkultur in Oldenburg, 1845–1945*. Edited by Ewald Gässler, the City of Oldenburg, and the Oldenburg City Museum. Veröffentlichungen des Stadtmuseums Oldenburg 5. Oldenburg: Isensee Verlag, 1989.

Bergerson, Andrew Stuart. "A History of Neighborliness in Alt-Hildesheim, 1900–1950: Custom, Transformation, Memory." PhD dissertation, University of Chicago, March 1998.

Bergerson, Andrew Stuart. *Ordinary Germans in Extraordinary Times: The Nazi Revolution in Hildesheim*. Bloomington: Indiana University Press, 2004.

Bergerson, Andrew Stuart. "Raum und Zeit in der deutschen Mittelstadt." Translated by Katrin Dalitz and Andrew S. Bergerson. In *Mittelstadt: urbanes Leben jenseits der Metropole*, edited by Brigitte Schmidt-Lauber, 245–60. Frankfurt: Campus Verlag, 2010.

Berman, Marshall. *All That Is Solid Melts into Air: The Experience of Modernity*. New York: Simon and Schuster, 1982.

Bernatzky, Aloys. *Von der mittelalterlichen Stadtbefestigung zu den Wallgrünflächen von Heute. Ein Beitrag zum Grünflächenproblem deutscher Städte*. Berlin: Bernhard Patzer-Verlag, 1960.

Bernet, Claus. "The 'Hobrecht Plan' (1862) and Berlin's Urban Structure." *Urban History* 31, no. 3 (2004): 402–8.

Bernhardt, Christoph. *Bauplatz Groß-Berlin: Wohnungsmärkte, Terraingewerbe und Kommunalpolitik im Städtewachstum der Hochindustrialisierung (1871–1918)*. Veröffentlichungen der Historischen Kommission zu Berlin 93. Berlin: Walter de Gruyter, 1998.

Bernhardt, Christoph. "'Soziale Mischung' mit begrenzter Wirkung—Eine empirische Prüfung der Hobrecht-These zur Wohnweise im Berliner Mietshaus des Kaiserreichs." In *Soziale Mischung in der Stadt: Case Studies—Wohnungspolitik in Europa—historische Analyse*, edited by Tilman Harlander, Gerd Kuhn, and Wüstenrot Stiftung, 48–53. Stuttgart: Krämer, 2012.

Beyer, Peter. "Leipzigs Auseinandersetzung mit Frankfurt am Main (1706–1726). Symptom des Aufstiegs zur führenden deutschen Messestadt." In *Leipzigs Messen, 1497–1997*, edited by Hartmut Zwahr, Thomas Topfstedt, and Günter Bentele, 193–204. Geschichte und Politik in Sachsen 1. Cologne: Böhlau Verlag, 1999.

Blackbourn, David. *The Conquest of Nature: Water, Landscape, and the Making of Modern Germany*. New York: W. W. Norton, 2006.

Blackbourn, David. "'The Garden of Our Hearts': Landscape, Nature, and Local Identity in the German East." In *Localism, Landscape, and the Ambiguities of Place: German-Speaking Central Europe, 1860–1930*, edited by James Retallack and David Blackbourn, 149–64. Toronto: University of Toronto Press, 2007.

Blackbourn, David. *History of Germany, 1780–1918: The Long Nineteenth Century*. Malden, MA: Blackwell, 2003.

Blackbourn, David, and James Retallack, eds. *Localism, Landscape, and the Ambiguities of Place: German-Speaking Central Europe, 1860–1930*. Toronto: University of Toronto Press, 2007.

Blaschke, Karlheinz. "Die kursächsische Politik und Leipzig im 18. Jahrhundert." In *Leipzig: Aufklärung und Bürgerlichkeit*, edited by Wolfgang Martens, 23–38. Zentren der Aufklärung 3. Heidelberg: Verlag Lambert Schneider, 1990.

Blessing, Werner K. "Der Schein der Provinzialität: Nürnberg im 19. Jahrhundert." In *Nürnberg: Eine Europäische Stadt in Mittelalter und Neuzeit*, edited by Helmut Neuhaus, 69–103. Nürnberger Forschungen 29. Nuremberg: Verein für Geschichte der Stadt Nürnberg, 2000.

Blickle, Peter. *Heimat: A Critical Theory of the German Idea of Homeland*. Rochester, NY: Camden House, 2002.

Bodenschatz, Harald. *Platz frei für das Neue Berlin! Geschichte der Stadterneuerung in der 'größten Mietskasernenstadt der Welt' seit 1871*. Berlin: Transit Verlag, 1987.

Boy, Hans. *Die Stadtlandschaft Oldenburg: Siedlungsgeographie einer Niedersächsischen Stadt*. Bremen-Horn: Walter Dorn Verlag, 1954.

Brantz, Dorothee. "Slaughter in the City: The Establishment of Public Abattoirs in Paris and Berlin, 1780–1914." PhD dissertation, University of Chicago, 2003.

Brantz, Dorothee, and Sonja Dümpelmann, eds. *Greening the City: Urban Landscapes in the Twentieth Century*. Charlottesville: University of Virginia Press, 2011.

Bräuer, Helmut. *Der Leipziger Rat und die Bettler: Quellen und Analysen zu Bettlern und Bettelwesen in der Messestadt bis ins 18. Jahrhundert*. Leipzig: Leipziger Universitätsverlag GmbH, 1997.

Braun, Hermann, and Michael Neumann. *Die Oldenburg Neustadtquartiere Dobben und Haarenesch*. Oldenburg: Heinz Holzberg Verlag, 1978.

Braun, Rudolf. "Taxation, Sociopolitical Structure, and State-Building: Great Britain and Brandenburg-Prussia." In *The Formation of National States in Western Europe*, edited by Charles Tilly, 243–327. Princeton, NJ: Princeton University Press, 1975.

Brechin, Gray. *Imperial San Francisco: Urban Power, Earthly Ruin*. 2nd ed. Berkeley: University of California Press, 2006.

Bredehorn, Georg. *Eversten: Von 1200 bis ins 20. Jahrhundert*. Oldenburg: Isensee Verlag, 2001.

Breuilly, John, and Iorwerth Prothero. "The Revolution as Urban Event: Hamburg and Lyon during the Revolutions of 1848–49." In *Europe in 1848: Revolution and Reform*, edited by Dieter Dowe, 371–400. Oxford: Berghahn Books, 2001.

Brian, Amanda M. "Art from the Gutter: Heinrich Zille's Berlin." *Central European History* 46, no. 1 (2013): 28–60.

Briel, Cornelia. *Innere Westvorstadt: Eine historische und städtebauliche Studie*. Leipzig: Pro Leipzig, 1998.

Brix, Michael. "Das Problem der Vorstädte und ihrer Anbindung an die innere Stadt in der Phase der Industrialisierung: Der Kampf um den Befestigungsring als Kulturdenkmal." In *Nürnberg und Lübeck im 19. Jahrhundert: Denkmalpflege, Stadtbildpflege, Stadtumbau*, 107–29. Munich: Prestel-Verlag, 1981.

Brockmann, Stephen. *Nuremberg: The Imaginary Capital*. Rochester, NY: Camden House, 2006.

Bruder, Thomas. *Nürnberg als bayerische Garnison von 1806 bis 1914: Städtebauliche, wirtschaftliche und soziale Einflüsse*. Edited by Michael Diefenbacher, Rudolf Endres and Gerhard Hirschmann. Nürnberger Werkstücke zur Stadt- und Landesgeschichte. Nuremberg: Schriftenreihe des Stadtarchivs Nürnberg, 1992.

Bryant, Chad. "Into an Uncertain Future: Railroads and Vormärz Liberalism in Brno, Vienna, and Prague." *Austrian History Yearbook* 40 (2009): 183–201.

Bühl-Gramer, Charlotte. *Nürnberg 1850 bis 1892: Stadtentwicklung, Kommunalpolitik und*

Stadtverwaltung im Zeichen von Industrialisierung und Urbanisierung. Nürnberger Werkstücke zur Stadt- und Landesgeschichte 62. Nuremberg: Schriftenreihe des Stadtarchivs Nürnberg, 2003.

Bullock, Nicholas, and James Read. *The Movement for Housing Reform in Germany and France, 1840–1914*. Cambridge Urban and Architectural Studies 9. Cambridge: Cambridge University Press, 1985.

Bünz, Enno. "Leipzig als landesherrliche Residenz vom 13. Bis 16. Jahrhundert." In *Leipzigs Bedeutung für die Geschichte Sachsens: Politik, Wirtschaft und Kultur in sechs Jahrhunderten*, edited by Detlev Döring, 23–48. Leipzig: Leipziger Universitätsverlag, 2014.

Casey, Edward S. *The Fate of Place: A Philosophical History*. Berkeley: University of California Press, 1997.

Çelik, Zeynep. *Empire, Architecture, and the City: French-Ottoman Encounters, 1830–1914*. Seattle: University of Washington Press, 2008.

Certeau, Michel de. *The Practice of Everyday Life*. Translated by Steven Rendall. Berkeley: University of California Press, 1984.

Certeau, Michel de. *The Writing of History*. Translated by Tom Conley. New York: Columbia University Press, 1988.

Ciancia, Kathryn. "Borderland Modernity: Poles, Jews, and Urban Spaces in Interwar Eastern Poland." *Journal of Modern History* 89 (September 2017): 531–61.

Clark, Catherine E. *Paris and the Cliché of History: The City and Photographs, 1860–1970*. Oxford: Oxford University Press, 2018.

Clausmeyer-Ewers, Bettina. *Die Wallanlagen in Frankfurt am Main: Entwicklung von der mittelalterlichen Stadtbefestigung zu den heutigen Wallgrünflächen*. Frankfurt am Main: Garten and Friedhofsamt, 1988.

Confino, Alon. *The Nation as a Local Metaphor: Württemberg, Imperial Germany, and National Memory, 1871–1918*. Chapel Hill: University of North Carolina Press, 1997.

Cosgrove, Denis. "Landscape and *Landschaft*: Lecture Delivered at the 'Spatial Turn in History' Symposium, German Historical Institute, February 19, 2004." *GHI Bulletin* 35 (Fall 2004): 57–71.

Cronon, William. *Nature's Metropolis: Chicago and the Great West*. New York: W. W. Norton, 1991.

Czok, Karl. *Vorstädte. Zu ihrer Entstehung, Wirtschaft und Sozialentwicklung in der älteren deutschen Stadtgeschichte*. Sitzungsberichte der Sächsischen Akademie der Wissenschaften zu Leipzig, Philogisch-historische Klasse 121, no. 1. [East] Berlin: Akademie-Verlag, 1979.

Danto, Arthur C. *Narration and Knowledge*. Introduction by Lydia Goehr; conclusion by Frank Ankersmit. New York: Columbia University Press, 2007.

Davis, Belinda. "'Everyday' Protest and the Culture of Conflict in Berlin, 1830–1980." In *Berlin—Washington, 1800–2000: Capital Cities, Cultural Representation, and National Identities*, edited by Andreas W. Daum and Christof Mauch, 263–84. Washington, DC: German Historical Institute, 2005.

Deuter, Jörg. *Oldenburg: ein norddeutsches Stadtbild*. Oldenburgische Monographien. Oldenburg: Heinz Holzberg Verlag, 1988.

Diamant, Adolf. *Chronik der Juden in Leipzig: Aufstieg, Vernichtung und Neuanfang*. Frankfurt am Main: Verlag Heimatland Sachsen, 1993.

Diekmann-Dröge, Gabriele. "Niederdeutsche Heimatliteratur in Oldenburg, 1870–1950." In *Suche nach Geborgenheit: Heimatbewegung in Stadt und Land Oldenburg*, edited by Uwe Meiners, 174–207. Oldenburg: Isensee Verlag, 2002.

Dierßen, Klaus. "Hochmoore aus biozönotischer und ökosystemarer Sicht—Konsequenzen und Erfolgsaussichten für deren Restitution." In *Moor—eine verlorene Landschaft: Vorträge anlässlich des Symposiums in Oldenburg vom 5.–7. November 1999*, edited by Carsten Ritzau and Mamoun Fansa, 30–42. Schriftenreihe des Landesmuseum für Natur und Mensch 20. Oldenburg: Isensee Verlag, 2001.

Dietrich, Jörg. *Panorama des Leipziger Rings 1850/2015*. Leipzig: Lehmstedt, 2015.

Di Palma, Vittoria. *Wasteland: A History*. New Haven, CT: Yale University Press, 2014.

Dollen, Busso von der. "Der Kampf um das Sterntor: Die Auseinandersetzungen um Abriß oder Erhaltung der letzten mittelalterlichen Torburg Bonns im 19. Jahrhundert." *Bonner Geschichtsblätter* 31 (1979): 83–121.

Dollen, Busso von der. "Stadtrandphänomene in historisch-geographischer Sicht." *Siedlungsforschung: Archäologie—Geschichte—Geographie* 1 (1983): 15–34.

Dubbini, Renzo. *Geography of the Gaze: Urban and Rural Vision in Early Modern Europe*. Translated by Lydia G. Cochrane. Chicago: University of Chicago Press, 2002.

Dumbsky, Walter. *Die deutschen Festungen von 1871 bis 1914: Strategische Bedeutung und technische Entwicklung*. Edited by Karl-Heinz Ruffmann and Hubert Rumpel. Erlanger Historische Studien 11. Frankfurt am Main: Peter Lang, 1987.

Eckhardt, Albrecht, and Heinrich Schmidt, eds. *Geschichte des Landes Oldenburg. Ein Handbuch*. Oldenburg: Heinz Holzberg Verlag, 1987.

Eddie, Scott M. "The Prussian Settlement Commission and Its Activities in the Land Market, 1886–1918." In *Germans, Poland, and Colonial Expansion to the East: 1850 through the Present*, edited by Robert L. Nelson, 39–63. New York: Palgrave Macmillan, 2009.

Emmrich, Julia. *Wachgeküsst! Die Erfindung des modernen Dortmund in den Stadtführern des 19. Jahrhunderts*. Cologne: SH-Verlag, 2005.

Endres, Rudolf, and Martina Fleischmann. *Nürnbergs Weg in die Moderne: Wirtschaft, Politik und Gesellschaft im 19. und 20. Jahrhundert*. Edited by Stadtsparkasse Nürnberg. Nuremberg: Verlag W. Tümmels, 1996.

Engeli, Christian. "Stadterweiterungen in Deutschland im neunzehnten Jahrhundert." In *Die Städte Mitteleuropas im neunzehnten Jahrhundert*, edited by Wilhelm Rausch, 47–72. Linz/Donau: Rudolf Trauner, 1983.

Erbe, Michael. "Berlin im Kaiserreich (1871–1918)." In *Geschichte Berlin, Zweiter Band: Von der Märzrevolution bis zur Gegenwart*, edited by Wilhelm Ribbe, 691–796. 3rd ed. Munich: Verlag C. H. Beck, 1987.

Erlin, Matt. *Berlin's Forgotten Future: City, History, and Enlightenment in Eighteenth-Century Germany*. Chapel Hill: University of North Carolina Press, 2004.

Escher, Felix. *Berlin und sein Umland: Zur Genese der Berliner Stadtlandschaft bis zum Beginn des 20. Jahrhunderts*. Einzelveröffentlichungen der Historischen Kommission zu Berlin, Publikation der Sektion für die Geschichte Berlins 47. Berlin: Colloquium Verlag, 1985.

Escher, Felix. "Stadtranderscheinungen in Berlin vom 17. bis zum frühen 19. Jahrhundert." *Siedlungsforschung: Archäologie—Geschichte—Geographie* 1 (1983): 87–101.

Evans, Richard J. *Death in Hamburg: Society and Politics in the Cholera Years, 1830–1910*. Oxford: Clarendon Press, 1987.

Fehl, Gerhard, and Juan Rodriguez-Lores, eds. *Stadterweiterungen, 1800–1875: Von den Anfängen des modernen Städtebaues in Deutschland*. Stadt Planung Geschichte 2. Hamburg: Hans Christians Verlag, 1983.

Fehl, Gerhard, and Juan Rodriguez-Lores, eds. *Stadt-Umbau: Die planmäßige Erneuerung europäischer Großstädte zwischen Wiener Kongreß und Weimarer Republik*, Stadt Planung Geschichte 17. Basel: Birkhäuser Verlag, 1995.

Fischer, Lothar. *Heinrich Zille in Selbstzeugnissen und Bilddokumenten*. Hamburg: Rowohlt, 1979.

Forsell, Håkan. *Property, Tenancy, and Urban Growth in Stockholm and Berlin, 1860–1920*. Translated by Paul Fischer. Edited by Richard Roger and Jean-Luc Pinol. Historical Urban Studies. Aldershot, UK: Ashgate, 2006.

Förster, Heinrich, Alwin Bielefeldt, and Walter Reinhold. *Zur Geschichte des deutschen Kleingartenwesens*. Frankfurt am Main: Union-Druckerei und Verlagsanstalt GmbH, 1931.

Franz, Kathrin. "'Leipzig steht ebenso im Rufe der Langweiligkeit seiner Ebene, wie der Reize seiner Promenaden.' Die Geschichte der Promenadenanlagen bis 1945." *Leipziger Blätter* 39 (2001): 86–88.

Friedl, Hans. *Biographisches Handbuch zur Geschichte des Landes Oldenburg*. Oldenburg: Isensee Verlag, 1992.

Friedrichs, Christopher R. *The Early Modern City, 1450–1750: A History of Urban Society in Europe*. Vol. 1. Edited by Robert Tittler. London: Longman, 1995.

Friedrichs, Christopher R. "How German Was the German Home Town?" *Central European History* 47, no. 3 (2014): 488–95.

Frisby, David. *Cityscapes of Modernity: Critical Explorations*. Malden, MA: Polity Press, 2001.

Fritzsche, Peter. *Reading Berlin 1900*. Cambridge, MA: Harvard University Press, 1996.

Fritzsche, Peter. *Stranded in the Present: Modern Time and the Melancholy of History*. Cambridge, MA: Harvard University Press, 2004.

Fuchs, Alois. *Paderborn*. Munich: Deutsche Kunstverlag, 1965.

Fürbringer, Leo. *Die Stadt Emden in Gegenwart und Vergangenheit*. Emden: Verlag von W. Schwalbe, 1892.

Gall, Lothar, ed. *Stadt und Bürgertum im 19. Jahrhundert*. Munich: R. Oldenbourg, 1990.

Gandy, Matthew. *The Fabric of Space: Water, Modernity, and the Urban Imagination*. Cambridge, MA: MIT Press, 2014.

Gäßler, Ewald, ed. *Klassizismus: Baukunst in Oldenburg, 1785–1860*. Oldenburg: Verlag Isensee, 1991.

Gäßler, Ewald. "Der Maler und Graphiker Bernhard Winter (1871–1964): Ein exponierter Vertreter und Förderer der Heimatbewegung in Oldenburg." In *Suche nach Geborgenheit: Heimatbewegung in Stadt und Land Oldenburg*, edited by Uwe Meiners, 136–73. Oldenburg: Isensee Verlag, 2002.

Gäßler, Ewald, ed. "Wenn Künstler heute ins Moor gehen: Das Moor in der Kunst nach 1945 bis in die Gegenwart." In *Moor—eine verlorene Landschaft: Vorträge anlässlich des Symposiums in Oldenburg vom 5.–7. November 1999*, edited by Carsten Ritzau and Mamoun Fansa, 117–33. Schriftenreihe des Landesmuseum für Natur und Mensch 20. Oldenburg: Isensee Verlag, 2001.

Gaumer, Eberhard. "Leben in Oldenburg—Leben mit Wasser: Ein Geographische Untersuchung zum Verhältnis von Wasserregime und Stadtentwicklung." Dissertation, Carl von Ossietzky Universität, Oldenburg, 1998.

Geist, Johann Friedrich, and Klaus Kürvers. *Das Berliner Mietshaus, 1740–1862: Ein dokumentarische Geschichte der 'von Wülcknitzschen Familienhäuser' vor dem Hamburger Tor, der Proletarisierung des Berliner Nordens und der Stadt im Übergang von der Residenz zu Metropole*. Munich: Prestel Verlag, 1980.

Geyler, Helmut. "Die Promenade: 150 Jahre Geschlossene Stadtringanlagen in Leipzig." *Leipziger Beobachter und Wochen-Programm* 13, no. 4 (April 1936): 41–44.

Giesecke, Albert. "Die Befestigungen der Stadt Nürnberg und Albrecht Dürer." *Der Burgwart: Zeitung für Wehrbau, Wohnbau und Städtebau* 22, no. 3/4 (1921): 25–38.

Gilly de Montaut, Wilhelm. *Festung und Garnison Oldenburg*. Oldenburg: Heinz Holberg Verlag, 1980.

Glatzer, Ruth. *Berlin wird Kaiserstadt: Panorama einer Metropole, 1871–1890*. Introduction by Lothar Gall. Berlin: Siedler Verlag, 1993.

Goff, Lisa. *Shantytown, USA: Forgotten Landscapes of the Working Poor*. Cambridge, MA: Harvard University Press, 2016.

Götz, Norbert. *Um Neugotik und Nürnberger Stil: Studien zum Problem der künstlerischen Vergangenheitsrezeption im Nürnberg des 19. Jahrhundert*. Edited by Verein für Geschichte der Stadt Nürnberg, Nürnberger Forschungen 23. Nuremberg: Selbstverlag des Vereins für Geschichte der Stadt Nürnberg, 1981.

Gröning, Gert. "Wollen Schweizer Kleingärtner wirklich etwas mit Schreber zu tun haben?" In *High & Low: Gärten zwischen Kunst, Luxus und Alltag*, edited by Annemarie Bücher and Johannes Stoffler, 12–21. Zürich: Topiaria Helvetica, 2013.

Gudermann, Rita. "Conviction and Constraint: Hydraulic Engineers and Agricultural Amelioration Projects in Nineteenth-Century Prussia." In *Germany's Nature*, edited by Thomas Lekan and Thomas Zeller, 33–54. New Brunswick, NJ: Rutgers University Press, 2005.

Gudermann, Rita. *Morastwelt und Paradies: Ökonomie und Ökologie in der Landwirtschaft am Beispiel der Meliorationen in Westfalen und Brandenburg (1830–1880)*. Forschungen zur Regionalgeschichte 35. Paderborn: Verlag Ferdinand Schöningh, 2000.

Guettel, Jens-Uwe. *German Expansionism, Imperial Liberalism and the United States, 1776–1945*. Cambridge: Cambridge University Press, 2012.

Gray, Marion W. "Prussia in Transition: Society and Politics under the Stein Reform Ministry of 1808." *Transactions of the American Philosophical Society* 76, no. 1 (1986): 1–175.

Green, Abigail. *Fatherlands: State-Building and Nationhood in Nineteenth-Century Germany*. Cambridge: Cambridge University Press, 2001.

Grobe, Peter. *Die Entfestigung Münchens (Kurzfassung)*. Miscellanea Bavarica Monacensia 27. Munich: Stadtarchiv, 1970.

Groeneveld, Sabine. "Far Away at Home in Qingdao (1897–1914)." *German Studies Review* 39, no. 1 (2016): 65–79.

Groß, Reiner. *Geschichte Sachsens*. Leipzig: Die Deutsche Bibliothek, 2001.

Gruber-Corr, Birgitt. *Stadterweiterung im Rheinland: Kommune, Bürger und Staat als Akteure im Entstehungsprozess der Bonner Südstadt, 1855–1890*. Veröffentlichungen des Stadtarchivs Bonn 64. Edited by Norbert Schlossmacher. Bonn: Stadt Bonn, 2004.

Gubser, Michael. *Time's Visible Surface: Alois Riegl and the Discourse on History and Temporality in Fin-de-Siècle Vienna*. Detroit: Wayne State University Press, 2006.

Haase, Carl. *Die Entstehung der westfälischen Städte*. 4th ed. Veröffentlichungen des Provinzialinstituts für Westfälische Landes- und Volksforschung des Landschaftsverbandes Westfalen-Lippe 1:11. Münster: Aschendorffsche Verlagsbuchhandlung, 1984.

Hagen, Joshua. "Architecture, Urban Planning, and Political Authority in Ludwig I's Munich." *Journal of Urban History* 35, no. 4 (2009): 459–85.

Hagen, Joshua. *Preservation, Tourism and Nationalism: The Jewel of the German Past*. Aldershot, UK: Ashgate, 2006.

Hahn, Anne. *Die Entfestigung der Stadt Saarlouis*. Schriften des Landkreises Saarlouis 4. Saint Ingbert: Röhrig Universitätsverlag, 2000.

Hahn, Hans-Werner. *Altständisches Bürgertum zwischen Beharrung und Wandel: Wetzlar 1689–1870*. Stadt und Bürgertum 2, edited by Lothar Gall. Munich: R. Oldenbourg, 1991.

Hahn, Hans-Werner. "Zunftproteste gegen den modernen Steuerstaat. Die Wetzlar Auseinandersetzungen um die Preußische Mahl- und Schlachtsteuer, 1820–1840." *Nassauische Annalen* 98 (1987): 173–98.

Hartog, Rudolf. *Stadterweiterungen im 19. Jahrhundert*. Schriftenreihe des Vereins zur Pflege kommunalwissenschaftlicher Aufgaben e.V. Berlin 6. Stuttgart: W. Kohlhammer GmbH, 1962.

Harvey, David. *Paris: Capital of Modernity*. New York: Routledge, 2006.

Haunfelder, Bernd, and Ute Olliges. *Die Promenade in Münster: Vom Festungsring zum Grüngürtel*. Münster: Aschendorffsche Verlagsbuchhandlung GmbH, 1994.

Heggen, Alfred. *Paderborn unter Preußischer Herrschaft, 1802–1806: Vorgeschichte und Verlauf der Säkularisation*. Paderborn Geschichte in Bildern, Dokumenten, Zeugnissen. Paderborn: Bonifatius Druckerei, 1988.

Heil, Peter. *Von der ländlichen Festungsstadt zur bürgerlichen Kleinstadt: Stadtumbau zwischen Deutschland und Frankreich: Landau, Haguenau, Selestat und Belfort zwischen 1871 und 1930*. Geschichtliche Landeskunde 49. Stuttgart: Franz Steiner Verlag, 1999.

Heinrich, Axel. "Gemalte Heimat—Der Raum Oldenburg." In *Suche nach Geborgenheit: Heimatbewegung in Stadt und Land Oldenburg*, edited by Uwe Meiners, 90–110. Oldenburg: Isensee Verlag, 2002.

Hengsbach, Arne. "Der Berliner Stadtrand um 1860." *Jahrbuch für Brandenburgische Landesgeschichte* 39 (1988): 144–45.

Hennebo, Dieter. *Entwicklung des Stadtgrüns von der Antike bis in die Zeit des Absolutismus*. Hannover: Patzer Verlag, 1979.

Herzog, Steffen. "Kursachsens Städte im 18. Jahrhundert—Materialien der Restaurationskommission von 1762/63 zum Städtewesen." In *Sachsen, 1763–1832: Zwischen Rétablissement und bürgerlichen Reformen*, edited by Uwe Schirmer, 101–8. Beucha: Sax-Verlag, 1996.

Hett, Benjamin Carter. *Death in the Tiergarten: Murder and Criminal Justice in the Kaiser's Berlin*. Cambridge, MA: Harvard University Press, 2004.

Hilliges, Marion. "Entfestigung: Planungskonzepte zur Urbanisierung der 'Leere' im 18. Jahrhundert." *Die alte Stadt* 13 (2004): 161–81.

Hinrichs, August. *Über Oldenburg*. Oldenburg: Heinz Holzberg Verlag, 1987.

Hinrichs, Ernst. "Oldenburg in der Zeit Herzog Peter Friedrich Ludwigs." In *Geschichte der Stadt Oldenburg Band 1: Von den Anfängen bis 1830*, edited by Ekkehard Seeber, 481–622. Oldenburg: Isensee Verlag, 1997.

Hirschmann, Gerhard. "Fortleben Reichsstädtischen Bewusstseins in Franken nach 1806?" In *Aus Sieben Jahrhunderten Nürnberger Stadtgeschichte: Ausgewählte Aufsätze von Gerhard Hirschmann zu Seinem 70. Geburtstag*, edited by Kuno Ulshöfer. Nürnberger Forschungen 25. Nuremberg: Selbstverlag des Vereins für Geschichte der Stadt Nürnberg, 1988.

Hoffmann-Axthelm, Dieter, and Ludovica Scarpa. *Berliner Mauern und Durchbrüche*. Berliner Topografien 7. Berlin: Verlag Ästhetik und Kommunication, 1987.

Hohmann, Klaus. *Bauten des Historismus in Paderborn, 1800–1920*. Studien und Quellen zur westfälischen Geschichte 28. Paderborn: Bonifatius, 1990.

Hosang, E. "Berliner Wohnungsnoth." *Über Land und Meer* 13, no. 46 (1872): 13–14.

Jackisch, Barry A. "The Nature of Berlin: Green Space and Visions of a New German Capital, 1900–1945." *Central European History* 47, no. 2 (2014): 307–33.

Jackson, John Brinckerhoff. *A Sense of Place, a Sense of Time*. New Haven, CT: Yale University Press, 1994.

Jefferies, Matthew. *Imperial Culture in Germany, 1871–1918*. New York: Palgrave Macmillan, 2003.

Jenkins, Jennifer. *Provincial Modernity: Local Culture and Liberal Politics in Fin-de-Siècle Hamburg*. Ithaca, NY: Cornell University Press, 2003.

Jerram, Leif. "Bureaucratic Passions and the Colonies of Modernity: An Urban Elite, City Frontiers and the Rural Other in Germany, 1890–1920." *Urban History* 34, no. 3 (2007): 390–405.

Jerram, Leif. *Germany's Other Modernity: Munich and the Building of Metropolis, 1895–1930*. Manchester: Manchester University Press, 2007.

Jindrich, Jason. "The Shantytowns of Central Park West: Fin de Siècle Squatting in American Cities." *Journal of Urban History* 36, no. 5 (2010): 672–84.

Johanek, Peter, ed. *Die Stadt und ihr Rand*. Cologne: Böhlau, 2008.

Jones, Elizabeth B. "Fixing Prussia's Peripheries: Rural Disasters and Prusso-German State-Building, 1866–1914." *Central European History* 51 (2018): 204–27.

Jordan, David P. *Transforming Paris: The Life and Labors of Baron Haussmann*. New York: Free Press, 1995.

Jung, Gernot, and Dirk Wundram. *Landschaftsgeschichte des Oldenburger Landes vom Ende das 19. bis zum Anfang des 21. Jahrhunderts*. Oldenburg: Isensee Verlag, 2006.

Kaeber, Ernst. "Das Weichbild der Stadt Berlin seit der Steinschen Städteordnung." In *Beiträge zur Berliner Geschichte: Ausgewählte Aufsätze*, edited by Werner Vogel, 234–376. Berlin: Walter de Gruyter, 1964.

Kaufhold, Enno. *Heinrich Zille: Photograph der Moderne*. Munich: Schirmer/Mosel, Berlinische Galerie, 1995.

Keinemann, Friedrich. *Von den Freiheitskriegen zur Julirevolution: Westfalen im frühen 19. Jahrhundert*. Norden: Verlag Hohes Riff, 2006.

Kevorkian, Tanya. "The Rise of the Poor, Weak, and Wicked: Poor Care, Punishment, Religion, and Patriarchy in Leipzig, 1700–1730." *Journal of Social History* 34, no. 1 (2000): 163–81.

Keyser, Erich, ed. *Deutsches Städtebuch: Handbuch städtischer Geschichte, Band II: Mitteldeutschland*. Stuttgart: W. Kohlhammer Verlag, 1941.

Kier, Hiltrud. "Die Stadterweiterungsplanung von Josef Stübben für die Kölner Neustadt ab 1880: Versuche städtebaulicher und sozialer Differenzierung mit dem Instrument der Fluchtlinienplanung." In *Städtebaureform, 1865–1900, Teil 1: Von Licht, Luft und Ordnung in der Stadt der Gründerzeit. Allgemeine Beiträge und Bebauungsplanung*, edited by Juan Rodriguez-Lores and Gerhard Fehl, 155–64. Hamburg: Hans Christians Verlag, 1985.

Kleinberg, Ethan, Joan Wallach Scott, and Gary Wilder. *Theses on Theory and History* (May 2018). https://versobooks.com/blogs/3893-theses-on-theory-and-history/.

Klemek, Christopher. *The Transatlantic Collapse of Urban Renewal: Postwar Urbanism from New York to Berlin*. Historical Studies of Urban America. Chicago: University of Chicago Press, 2011.

Koch, Herbert. *Geschichte der Stadt Jena*. Stuttgart: Gustav Fischer Verlag, 1966.

Kooij, Pam, and Piet Pellenbarg. *Regional Capitals: Past, Present, Prospects*. Assen, Netherlands: Van Gorcum, 1994.

Kopp, Kristin. *Germany's Wild East: Constructing Poland as Colonial Space.* Ann Arbor: University of Michigan Press, 2012.

Koselleck, Reinhart. *Futures Past: on the Semantics of Historical Time.* Translated and with an introduction by Keith Tribe. New York: Columbia University Press, 2004.

Koshar, Rudy. "Against the 'Frightful Leveler': Historic Preservation and German Cities, 1890–1914." *Journal of Urban History* 19, no. 3 (1993): 7–29.

Koshar, Rudy. *From Monuments to Traces: Artifacts of German Memory, 1870–1990.* Berkeley: University of California Press, 2000.

Koshar, Rudy. *German Travel Cultures.* Oxford: Berg, 2000.

Koshar, Rudy. *Germany's Transient Pasts: Preservation and National Memory in the Twentieth Century.* Chapel Hill: University of North Carolina Press, 1998.

Krabbe, Wolfgang R. *Die deutsche Stadt im 19. und 20. Jahrhundert. Eine Einführung.* Göttingen: Vandenhoeck and Ruprecht, 1989.

Krüger, Kersten. "Wandel des Stadtbildes durch Festungsbau—Oldenburg in Dänischer Zeit." *Oldenburger Jahrbuch* 87 (1987): 47–108.

Krüger, Ulrich. *Carl Heine: Der Mann, der Leipzig zur Industriestadt machte.* Erfurt: Sutton Verlag, 2008.

Kühn, Christoph. *Ostvorstadt: Eine historische und städtebauliche Studie.* Leipzig: Pro-Leipzig, 1998.

Kuhn, Gerd. "Durcheinander Wohnen und Zonierung in Kaiserreich." In *Soziale Mischung in der Stadt: Case Studies—Wohnungspolitik in Europa—historische Analyse*, edited by Tilman Harlander, Gerd Kuhn, and Wüstenrot Stiftung, 28–47. Stuttgart: Krämer, 2012.

Kusch, Eugen. *Nürnberg: Lebensbild einer Stadt.* Nuremberg: Verlag Nürnberger Presse, 1989.

Küster, Bernd. "Die Malerei der Moore." In *Moor—eine verlorene Landschaft: Vorträge anlässlich des Symposiums in Oldenburg vom 5.–7. November 1999*, edited by Carsten Ritzau and Mamoun Fansa, 107–15. Schriftenreihe des Landesmuseum für Natur und Mensch 20. Oldenburg: Isensee Verlag, 2001.

Ladd, Brian. "The Closed versus the Open Cityscape: Rival Traditions from Nineteenth-Century Europe." *Change Over Time* 4, no. 1 (2014): 58–74.

Ladd, Brian. *The Ghosts of Berlin: Confronting German History in the Urban Landscape.* Chicago: University of Chicago Press, 1997.

Ladd, Brian. *Urban Planning and Civic Order in Germany, 1860–1914.* Harvard Historical Studies. Cambridge, MA: Harvard University Press, 1990.

Lampe, Klaus. "Wirtschaft und Gesellschaft Oldenburgs um 1800." In *Peter Friedrich Ludwig und das Herzogtum Oldenburg. Beiträge zur Oldenburgischen Landesgeschichte um 1800*, edited by Heinrich Schmidt, 15–42. Oldenburg: Heinz Holzberg Verlag, 1979.

Lange, Annemarie. *Berlin zur Zeit Bebels und Bismarcks: Zwischen Reichsgründung und Jahrhundertwende.* Berlin: Dietz Verlag, 1972.

Large, David Clay. *Berlin.* New York: Basic Books, 2000.

Larson, Susan. *Constructing and Resisting Modernity: Madrid, 1900–1936*. Madrid: Iberoamericana, 2011.

Lees, Andrew. "Between Anxiety and Admiration: Views of British Cities in Germany, 1835–1914." *Urban History* 36, no. 1 (2009): 42–66

Lees, Andrew. *Cities Perceived: Urban Society in European and American Thought, 1820–1940*. Manchester: Manchester University Press, 1985.

Lees, Andrew. *Cities, Sin, and Social Reform in Imperial Germany*. Ann Arbor: University of Michigan Press, 2002.

Lees, Andrew. "The Civic Pride of the German Middle Classes, 1890–1918." In *Another Germany: A Reconsideration of the Imperial Era*. Edited by Jack Dukes and Joachim Remak, 41–59. Boulder, CO: Westview, 1987.

Lees, Andrew, and Lynn Hollen Lees. *Cities and the Making of Modern Europe, 1750–1914*. Cambridge: Cambridge University Press, 2007.

Leesch, Wolfgang. *Geschichte der Steuerverfassung und -verwaltung in Westfalen seit 1815*. Paderborn: Verlag Bonifatius-Druckerei, 1983.

Lehmann, Pit. "Vom Verteidigungsgelände zum Promenadenring." In *Leipzig um 1800: Beiträge zur Sozial- und Kulturgeschichte*, edited by Thomas Topfstedt and Hartmut Zwahr, 7–15. Beucha: Sax-Verlag, 1998.

Lekan, Thomas M. *Imagining the Nation in Nature: Landscape Preservation and German Identity, 1885–1945*. Cambridge, MA: Harvard University Press, 2004.

Lempa, Heikki. *Beyond the Gymnasium: Educating the Middle-Class Bodies in Classical Germany*. Lanham, MD: Lexington Books, 2007.

Lenger, Friedrich. *Sozialwissenschaft um 1900: Studien zu Werner Sombart und einigen seiner Zeitgenossen*. Frankfurt am Main: Peter Lang, 2009.

Lenger, Friedrich, ed. *Towards an Urban Nation: Germany since 1780*. German Historical Perspectives 16. Oxford: Berg, 2002.

Lenger, Friedrich. *Zwischen Kleinbürgertum und Proletariat: Studien zur Sozialgeschichte der Düsseldorfer Handwerker 1816–1878*. Kritische Studien zur Geschichtswissenschaft 71. Göttingen: Vandenhoeck and Ruprecht, 1986.

Lenman, Robin. *Artists and Society in Germany, 1850–1914*. Manchester: Manchester University Press, 1997.

Limann, Georg. "Hydrographie der Stadt Oldenburg." *Oldenburger Jahrbuch* 52 (1952/53): 140–59.

Lindemann, Mary. *Patriots and Paupers: Hamburg, 1712–1830*. New York: Oxford University Press, 1990.

Lonner, Alyssa. "History's Attic: Artifacts, Museums, and Historical Rupture in Gustav Freytag's *Die verlorene Handschrift*." *Germanic Review* 82, no. 4 (September 2007): 321–42.

Lüdtke, Alf. *Police and State in Prussia, 1815–1850*. Translated by Pete Burgess. Cambridge: Cambridge University Press, 1989.

Ludwig, Jörg. "Die Leipziger Messe in den ersten Jahrzehnten des Zollvereins (1834–1860)." In *Leipzigs Messen, Teilband 1: 1497–1997*, edited by Hartmut Zwahr, Thomas Topfstedt, and Günter Bentele, 351–65. Cologne: Böhlau Verlag, 1999.

Lutz, Leonhard. *Gesellschaft und Wirtschaft Nürnbergs unter Otto Freiherr von Stromer*. Edited by Rainer Gömmel and Jürgen Schneider. Beiträge zur Wirtschafts- und Sozialgeschichte 52. Stuttgart: Franz Steiner Verlag, 1992.

Maderthaner, Wolfgang. "Outcast Vienna 1900: The Politics of Transgression." In *Spaces of the Poor*, edited by Hans-Christian Petersen, 121–34. Bielefeld: Transcript, 2013.

Maderthaner, Wolfgang. *Unruly Masses: The Other Side of Fin-de-Siècle Vienna*. New York: Berghahn Books, 2008.

Maier, Charles. *Once within Borders: Territories of Power, Wealth, and Belonging since 1500*. Cambridge, MA: Belknap Press of Harvard University Press, 2016.

Maron, Wolfgang. "Vom Ende des Fürstbistums bis zur Gründung des Deutschen Reiches (1802–1871)." In *Paderborn: Geschichte der Stadt in ihrer Region Band 3: Das neunzehnte und zwanzigste Jahrhundert: Traditionsbindung und Modernisierung*, edited by Karl Hüser, 3–99. Paderborn: Ferdinand Schöningh, 1999.

Masur, Gerhard. *Imperial Berlin*. New York: Basic Books, 1970.

Mattenklott, Gert, and Gundel Mattenklott. *Berlin Transit. Eine Stadt als Station*. Reinbek bei Hamburg: Rowohlt, 1987.

McNeur, Catherine. *Taming Manhattan: Environmental Battles in the Antebellum City*. Cambridge, MA: Harvard University Press, 2014.

Meckseper, Cord, and Harald Siebenmorgen, eds. *Die alte Stadt: Denkmal oder Lebensraum? Die Sicht der mittelalterlichen Stadtarchitektur im 19. und 20. Jahrhundert*. Göttingen: Vandenhoeck and Ruprecht, 1985.

Meller, Helen. "Citizens in Pursuit of Nature: Gardens, Allotments, and Private Space in European Cities, 1850–2000." In *Resources of the City: Contributions to an Environmental History of Modern Europe*, edited by Dieter Schott, Bill Luckin, and Geneviève Massard-Guilbaud, 80–96. Aldershot, UK: Ashgate, 2005.

Menne, Paul. *Die Festungen des norddeutschen Raumes*. Schriften der Wirtschaftswissenschaftlichen Gesellschaft zum Studium Niedersachsens E.V., Neue Folge 18. Oldenburg: Verlag Gerhard Stalling, 1942.

Menninger, Margaret. "Art and Civic Patronage in Leipzig, 1848–1914." PhD dissertation, Harvard University, Cambridge, Massachusetts, 1998.

Merriman, John H. *The Margins of City Life: Explorations on the French Urban Frontier, 1815–1851*. New York: Oxford University Press, 1991.

Metscher, Karl, and Walter Fellmann. *Lipsia und Merkur: Leipzig und seine Messen*. Leipzig: VEB F. A. Brockhaus Verlag, 1990.

Mettele, Gisela. *Bürgertum in Köln, 1775–1870: Gemeinsinn und freie Association*. Stadt und Bürgertum. Munich: R. Oldenburg Wissenschaftsverlag, 1998.

Mettele, Gisela. "Burgher Cities on the Road to a Civil Society: Germany 1780–1870." In *Towards an Urban Nation: Germany since 1780*, edited by Friedrich Lenger, 43–68. German Historical Perspectives 16. Oxford: Berg, 2002.

Meyer, Heinz. "Natürräumliche Bedingungen für die Entwicklung Osternburg." In *Osternburg: ein Ort mit vielen Gesichtern*, edited by Matthias Schachtschneider, 13–22. Oldenburg: Isensee Verlag, 1999.

Meyer, Karl Otto, ed. *Oldenburg und das Wasser. Beiträge aus der Geschichte der Kartographie.* Oldenburger Beiträge zur Landeskunde 1. Oldenburg: Isensee Verlag, 1990.

Meyer, Lioba. "Das Industrie- und Arbeiterviertel in Osternburg." In *Osternburg: Ein Ort mit vielen Gesichtern*, edited by Matthias Schachtschneider, 89–129. Oldenburg: Isensee Verlag, 1999.

Mezzadra, Sandro, "Zur Theorie der Stadt bei Hugo Preuß und Max Weber." In *Geschichte, State, Staat: Aspekte der Verfassungstheorie von Hugo Preuß*, edited by Christoph Müller, 85–99. Baden-Baden: Nomos Verlagsgesellschaft, 2005.

Mintzker, Yair. *The Defortification of the German City, 1689–1866*. Cambridge: Cambridge University Press, 2012.

Mintzker, Yair. "The Paradox of Visual and Material Cultures in Mack Walker's German Home Towns." *Central European History* 47, no. 3 (2014): 505–12.

Moor, Martina De, Leigh Shaw-Taylor and Paul Warde. "Comparing the Historical Commons of North West Europe. An Introduction." In *The Management of Common Land in North West Europe, c. 1500–1850*, 15–31. Turnhout, Belgium: Brepols, 2002.

Muller, Jurgen. "Crumbling Walls: Urban Change in Eighteenth-Century Germany." *German Studies Review* 19, no. 2 (1996): 225–39.

Nelson, Robert, ed. *Germans, Poland, and Colonial Expansion to the East: 1850 through the Present*. Studies in European Culture and History, edited by Eric D. Weitz and Jack Zipes. New York: Palgrave Macmillan, 2009.

Neumann, Michael. *Stadtplanung und Wohnhausbau in Oldenburg, 1850–1914*. Oldenburger Studien 22. Oldenburg: Heinz Holzberg, 1982.

Neumeyer, Eva Maria. "The Landscape Garden as a Symbol in Rousseau, Goethe and Flaubert." *Journal of the History of Ideas* 8, no. 2 (1947): 187–217.

Newhouse, Amy Melinda. "Outside the Walls: Civic Belonging and Contagious Disease in Sixteenth-Century Nuremberg." PhD dissertation, University of Arizona, Tucson, 2017.

Nilsen, Micheline. *The Working Man's Green Space: Allotment Gardens in England, France, and Germany, 1870–1919*. Charlottesville: University of Virginia Press, 2014.

Nipperdey, Thomas. *Nachdenken über die deutsche Geschichte: Essays*. Munich: Verlag C. H. Beck, 1986.

Nistal, Matthias. "Oldenburg wird moderne Hauptstadt. Vom Ende des Ersten Weltkrieges bis zum Beginn des Nationalsozialismus (1918–1932)." In *Geschichte der Stadt Oldenburg Band 2: 1830–1995*, edited by Ekkehard Seeber, 287–390. Oldenburg: Isensee Verlag, 1996.

Nitsche, Rainer, ed. *Häuserkämpfe, 1872, 1920, 1945, 1982*. Berlin: Transit, 1981.

Novy, Klaus, and Michael Prinz. *Illustrierte Geschichte der Gemeinwirtschaft: Wirtschaftliche Selbsthilfe in der Arbeiterbewegung von den Anfängen bis 1945*. Berlin: Verlag J. H. W. Dietz Nachf., 1985.

Obitz, Günther. *Geschichten vom Wall*. Illustrated by Helge-Marie Wandel. Bremen: H. M. Hauschild, 1994.

O'Donnell, Krista. "Home, Nation, Empire: Domestic Germanness and Colonial Citizenship." In *The Heimat Abroad: The Boundaries of Germanness*, edited by Krista O'Donnell, Renate Bridenthal, and Nancy Reagin, 40–57. Ann Arbor: University of Michigan Press, 2005.

Oettermann, Stephen. *The Panorama: History of a Mass Medium*. Translated by Deborah Lucas Schneider. New York: Zone Books, 1997.

Osterhammel, Jürgen. *Die Verwandlung der Welt: Eine Geschichte des 19. Jahrhunderts*. Munich: C. H. Beck, 2009.

Palmowski, Jan. *Urban Liberalism in Imperial Germany: Frankfurt am Main, 1866–1914*. Oxford: Oxford University Press, 1999.

Papke, Eva. *Festung Dresden: Aus der Geschichte der Dresdner Stadtbefestigung*. Edited by Staatliche Schlösser und Gärten Dresden. Dresden: Michel Sandstein Verlagsgesellschaft, 1997.

Parisius, Bernhard. *Vom Groll der 'kleinen Leute' zum Programm der kleinen Schritte: Arbeiterbewegung im Herzogtum Oldenburg, 1840–1890*. Oldenburg: Heinz Holzberg Verlag, 1985.

Pelc, Ortwin. *Im Schutz von Mauern und Toren: Die Befestigung der schleswig-holsteinischen Städte in Mittelalter und Neuzeit*. Heide: Boyens, 2003.

Penny, H. Glenn. "Fashioning Local Identities in an Age of Nation-Building: Museums, Cosmopolitan Visions, and Intra-German Competition." *German History* 17, no. 4 (1999): 489–505.

Penny, H. Glenn. *Kindred by Choice: Germans and American Indians since* 1800. Chapel Hill: University of North Carolina Press, 2013.

Pingel, Norman-Mathias. *Stadterweiterung und städtische Behörden in Braunschweig, 1851–1914*. Braunschweiger Werkstücke 99. Veröffentlichungen aus dem Stadtarchiv und der Stadtbibliothek 45a. Hannover: Verlag Hahnsche Buchhandlung, 1998.

Poling, Kristin. "Shantytowns and Pioneers beyond the City Wall: Berlin's Urban Frontier in the Nineteenth Century." *Central European History* 47, no. 2 (2014): 245–74.

Pollak, Martha. *Cities at War in Early Modern Europe*. Cambridge: Cambridge University Press, 2010.

Prestel, Joseph Ben. *Emotional Cities: Debates on Urban Change in Berlin and Cairo, 1860–1910*. Oxford: Oxford University Press, 2017.

Pritzl, Hans. *Mainz im Wandel zu einer offenen Stadt: Der Mainzer Grüngürtel*. Inaugural-Dissertation zur Erlangung des Grades einer Doktors der Technikwissenschaften, vorgelegt der Architektur-Fakultät der Erzherzog Johann Universität, 1993.

Prokopovych, Markian. *Habsburg Lemberg: Architecture, Public Space, and Politics in the Galician Capital, 1772–1914*. West Lafayette, IN: Purdue University Press, 2009.

Ranke, Winfried. *Vom Milljöh ins Milieu: Heinrich Zilles Aufstieg in der Berliner Gesellschaft.* Hannover: Fackelträger Verlag, 1979.

Rausch, Wilhelm, ed. *Die Städte Mitteleuropas im 19. Jahrhundert.* Linz: Druckerei Rudolf Trauner, 1983.

Reagin, Nancy R. "German *Brigadoon?* Domesticity and Metropolitan Germans' Perceptions of *Auslandsdeutschen* in Southwest Africa and Eastern Europe." In *The Heimat Abroad: The Boundaries of Germanness*, edited by Krista O'Donnell, Renate Bridenthal, and Nancy Reagin, 248–86. Ann Arbor: University of Michigan Press, 2005.

Reck, Hans-Hermann. *Die Stadterweiterung Triers: Planung und Baugeschichte von Beginn der preußischen Zeit bis zum Ende des Ersten Weltkrieges (1815–1918).* Trier: Verlag Trierer Historischer Forschungen, 1990.

Reinborn, Dietmar. *Städtebau im 19. und 20. Jahrhundert.* Stuttgart: Verlag W. Kohlhammer, 1996.

Reinders-Düselder, Christoph. "Oldenburg in 19. Jahrhundert—Auf dem Weg zur Selbstverwalteten Stadt 1830–1880." In *Geschichte der Stadt Oldenburg Band 2: 1830–1995*, edited by Ekkehard Seeber, 9–172. Oldenburg: Isensee Verlag, 1996.

Reinhold, Josef. *Zwischen Aufbruch und Beharrung: Juden und jüdische Gemeinde in Leipzig während des 19. Jahrhunderts.* Dresden: Sächsische Druck- und Verlagshaus AG, 1999.

Remak, Joachim, and Jack R. Dukes, eds. *Another Germany: A Reconsideration of the Imperial Era.* Boulder, CO: Westview, 1987.

Repp, Kevin. *Reformers, Critics, and the Paths of German Modernity: Anti-Politics and the Search for Alternatives, 1890–1914.* Cambridge, MA: Harvard University Press, 2000.

Retallack, James, ed. *Saxony in German History: Culture, Society, and Politics, 1830–1933.* Ann Arbor: University of Michigan Press, 2000.

Reulecke, Jürgen. *Geschichte der Urbanisierung in Deutschland.* Frankfurt am Main: Suhrkamp, 1985.

Ribhegge, Wilhelm. *Stadt und Nation in Deutschland vom Mittelalter bis zur Gegenwart: Die Entstehung der Zivilgesellschaft aus der Tradition der Städte.* Münster: Waxmann, 2002.

Richie, Alexandra. *Faust's Metropolis: A History of Berlin.* New York: Carroll and Graf, 1988.

Rodgers, Daniel T. *Atlantic Crossings: Social Politics in a Progressive Age.* Cambridge, MA: Belknap Press of Harvard University Press, 1998.

Rollins, William H. *A Greener Vision of Home: Cultural Politics and Environmental Reform in the German Heimatschutz Movement, 1904–1918.* Ann Arbor: University of Michigan Press, 1996.

Rönnebeck, Thomas. *Stadterweiterung und Verkehr im neunzehnten Jahrhundert.* Schriftenreihe der Institute für Städtebau der Technischen Hochschulen und Universitäten 5. Stuttgart: K. Krämer, 1971.

Rosenstrauch, Hazel. "Leipzig als 'Centralplatz' des deutschen Buchhandels." In *Leipzig: Aufklärung und Bürgerlichkeit*, edited by Wolfgang Martens, 103–24. Heidelberg: Verlag Lambert Schneider, 1990.

Ross, Anna. "Down with the Walls! The Politics of Place in Spanish and German Urban Extension Planning, 1848–1914." *Journal of Modern History* 90 (June 2018): 292–322.

Rubin, Eli. *Amnesiopolis: Modernity, Space, and Memory in East Germany*. Oxford: Oxford University Press, 2016.

Sahlins, Peter. *Boundaries: The Making of France and Spain in the Pyrenees*. Berkeley: University of California Press, 1989.

Saldern, Adelheid von. *Häuserleben zur Geschichte städtischen Arbeiterwohnens vom Kaiserreich bis heute*. Bonn: J. H. W. Dietz Nachfolger, 1995.

Schachtschneider, Matthias, ed. *Osternburg: Ein Ort mit vielen Gesichtern*. Oldenburg: Isensee Verlag, 1999.

Schaer, Friedrich-Wilhelm, and Albrecht Eckhardt. "Herzogtum und Großherzogtum Oldenburg im Zeitalter des aufgeklärten Absolutismus (1773–1847)." In *Geschichte des Landes Oldenburg: Ein Handbuch*, edited by Albrecht Eckhardt and Heinrich Schmidt. Oldenburg: Heinz Holzberg Verlag, 1996.

Scheffler, Jürgen. "'Weltstadt' und 'Unterwelt.' Urbanisierung, Armenpolitik und Obdachlosigkeit in Berlin 1871–1914." *Internationale Wissenschaftliche Korrespondenzen zur Geschichte der Deutschen Arbeiterbewegung* 22, no. 2 (June 1990), 158–81.

Schivelbusch, Wolfgang. *The Railway Journey: The Industrialization of Time and Space in the Nineteenth-Century*. Berkeley: University of California Press, 1986.

Schlör, Joachim. *Nights in the Big City: Paris, Berlin, London, 1840–1930*. Translated by Pierre Gottfried Imhof and Dafydd Rees Roberts. London: Reaktion Books, 1998.

Schlüpfinger, Heinrich. *Schwabach: Stadtgeschichte und Straßenlexikon in Wort und Bild*. Schwabach: Verlag Hermann, Millizer, 1989.

Schmidt, Heinrich. "Eversten—80 Jahre Stadtteil von Oldenburg: Betrachtungen zur Geschichte einer Wechselbeziehung, Vortrag beim Bürgerverein am 5. Oktober 2004." In *Eversten: Oldenburger Ansichten*, edited by Hans-Günther Zemke, 10–12. Oldenburg: Isensee Verlag, 2005.

Schmidt, Heinrich. "Herzog Peter Friedrich Ludwig von Oldenburg—Dynastische Voraussetzungen und wichtigste Lebensdaten." In Schmidt, *Peter Friedrich Ludwig*, 9–14.

Schmidt, Heinrich. "Oldenburg in Mittelalter und Früher Neuzeit." In Heinrich Schmidt and Ernst Hinrichs, *Geschichte der Stadt Oldenburg Band 1: Von den Anfängen bis 1830*, 11–477, edited by Ekkehard Seeber. Oldenburg: Isensee Verlag, 1997.

Schmidt, Heinrich, ed. *Peter Friedrich Ludwig und das Herzogtum Oldenburg. Beiträge zur Oldenburgischen Landesgeschichte um 1800*. Oldenburg: Heinz Holzberg Verlag, 1979.

Schmidt, Heinrich, Ernst Hinrichs, and Udo Elerd. *Geschichte der Stadt Oldenburg Band 1: Von den Anfängen bis 1830*. Edited by Ekkehard Seeber. Oldenburg: Isensee Verlag, 1997.

Schmidt, Herbert. "125 Jahre Eisenbahn im Landkreis Oldenburg." In *Der Landkreis Oldenburg: Menschen—Geschichte—Landschaft*, edited by Walter Barton and Wolfgang Haubold, 539–56. Oldenburg: Heinz Holzberg Verlag, 1992.

Schmidt, Jürgen. *Begrenzte Spielräume: Ein Beziehungsgeschichte von Arbeiterschaft und Bürgertum am Beispiel Erfurts, 1870–1914*. Göttingen: Vandenhoeck and Ruprecht, 2005.

Schmuhl, Hans-Walter. *Die Herren der Stadt: Bürgerliche Eliten und städtische Selbstverwaltung in Nürnberg und Braunschweig vom 18. Jahrhundert bis 1918*. Gießen: Focus Verlag, 1998.

Schneider, Wolfgang. *Leipzig: Streifzüge durch die Kulturgeschichte*. Leipzig: Gustav Kiepenheuer Verlag, 1995.

Schön, Bettina. "Die Entstehung des Promenadenrings." In *Der Leipziger Augustsplatz: Funktionen und Gestaltwandel eines Großstadtplatzes*, edited by Thomas Topfstedt and Pit Lehmann, 11–16. Leipzig: Leipziger Universitätsverlag, 1994.

Schönbach, Eva-Maria. "Preußische Verwaltung, politische Umbruch und die Anfänge der Moderne (1787–1847)." In *Geschichte der Stadt und Region Hamm im 19. und 20. Jahrhundert*, edited by Wilhelm Ribhegge, 11–71. Düsseldorf: Schwann im Patmos Verlag, 1991.

Schorske, Carl E. *Fin-de-Siècle Vienna: Politics and Culture*. 1st Vintage Book edition. New York: Vintage Books, 1981.

Schott, Dieter, Bill Luckin, and Geneviève Massard-Guilbaud, eds. *Resources of the City: Contributions to an Environmental History of Modern Europe*. Aldershot, UK: Ashgate, 2005.

Schröder, Julia H. "Großstadtklang: Hinterhofmusik als Teil urbaner Klanglandschaft. Die Harfenjule in Berlin um 1900." *Archiv für Musikwissenschaft* 69, no. 4 (2012): 349–62.

Schulze, Heinz-Joachim. *Oldenburgs Wirtschaft Einst und Jetzt*. Oldenburg: Gewerb- und Handelsverein von 1840, 1965.

Schwartz, Vanessa R. *Spectacular Realities: Early Mass Culture in Fin-de-Siècle Paris*. Berkeley: University of California Press, 1998.

Schwela, Christian. "Zentrum und Peripherie. Stadtkern, Vorstädte und Umland von Leipzig im Frühen 19. Jahrhundert." In *Leipzig um 1800: Beiträge zur Sozial- und Kulturgeschichte*, edited by Thomas Topfstedt, Hartmut Zwahr, and Katrin Löffler, 163–72. Beucha: Sax-Verlag, 1998.

Schwemmer, Wilhelm. *Die Stadtmauer von Nürnberg*. Führer zu Grossen Baudenkmälern 31. Berlin: Deutscher Kunstverlag, 1944.

Schwemmer, Wilhelm. "Die Stadtmauer von Nürnberg: Verluste und Erhaltung im 19. und 20. Jahrhundert." *Mitteilungen des Vereins für Geschichte der Stadt Nürnberg* 56 (1969): 424–44.

Schwibbe, Gudrun. *Wahrgenommen: Die Sinnliche Erfahrung der Stadt*. Münster: Waxmann, 2002.

Scott, James C. *Seeing like a State: How Certain Schemes to Improve the Human Condition Have Failed*. 3rd ed. New Haven, CT: Yale University Press, 1998.

Seggern, Andreas von. *Großstadt wider Willen: Zur Geschichte der Aufnahme und Integration von Flüchtlingen und Vertriebenen in der Stadt Oldenburg nach 1944*. Fremde Nähe— Beiträge zur interkulturellen Diskussion 8. Edited by Reimer Gronemeyer and Roland Schopf. Münster: LIT, 1997.

Sennett, Richard. *Conscience of the Eye: The Design and Social Life of Cities*. New York: Alfred A. Knopf, 1990.

Shapiro, Ann-Louis. *Housing the Poor of Paris, 1850–1902*. Madison: University of Wisconsin Press, 1985.

Sheehan, James. "Liberalism and the City in Nineteenth-Century Germany." *Past and Present* 51 (1971): 116–37.

Siegert, Rosemarie. *Steuerpolitik und Gesellschaft: Vergleichende Untersuchungen zu Preußen und Baden 1815–1848*. Edited by Wolfram Fischer. Schriften zur Wirtschafts- und Sozialgeschichte 63. Berlin: Duncker and Humblot, 2001.

Slotkin, Richard. *The Fatal Environment: The Myth of the Frontier in the Age of Industrialization, 1800–1890*. New York: Atheneum, 1985.

Smith, Helmut Walser. *The Continuities of German History: Nation, Religion, and Race across the Long Nineteenth Century*. Cambridge: Cambridge University Press, 2008.

Smith, Jeffrey Chipps. "Nuremberg and the Topographies of Expectation." *Journal of the Northern Renaissance* 1 (2009): 2. http://northernrenaissance.org/nuremberg-and-the-topographies-of-expectation/.

Smith, Woodruff D. "Friedrich Ratzel and the Origins of Lebensraum." *German Studies Review* 3, no. 1 (February 1980): 51–68.

Spaulding, Robert Mark. "Revolutionary France and the Transformation of the Rhine." *Central European History* 44, no. 2 (2011): 203–26.

Sperber, Jonathan. "Competing Counterrevolutions: Prussian State and Catholic Church in Westphalia during the 1850s." *Central European History* 19, no. 1 (1986): 45–62.

Spoerer, Mark. *Steuerlast, Steuerinzidenz, und Steuerwettbewerb: Verteilungswirkungen der Besteuerung in Preußen und Württemberg (1815–1913)*. Edited by Reinhard Spree. Jahrbuch für Wirtschaftsgeschichte 6. Berlin: Akademie Verlag, 2004.

Stein, Günter. "Berlins Stadtmauer." *Jahrbuch für brandenburgische Landesgeschichte* 6 (1955): 1–3.

Stein, Hartwig. *Inseln im Häusermeer: Eine Kulturgeschichte des deutschen Kleingartenwesens bis zum Ende des Zweiten Weltkrieges*. 2nd ed. Frankfurt am Main: Peter Lang, 2000.

Sternberger, Dolf. *Panorama of the Nineteenth Century*. Translated by Joachim Neugroschel. New York: Urizen Books Mole Editions, 1977.

Strauss, Gerald. "The Holy Roman Empire Revisited." *Central European History* 11, no. 2 (September 1978): 290–301.

Strauss, Gerald. *Nuremberg in the Sixteenth Century: City Politics and Life between the Middle Ages and Modern Times*. 2nd ed. Bloomington: Indiana University Press, 1976.

Strohmeyer, Klaus. *James Hobrecht (1825–1902) und die Modernisierung der Stadt*. Potsdam: Verlag für Berlin-Brandenburg, 2000.

Sutcliffe, Anthony. *Towards the Planned City: Germany, Britain, the United States and France, 1780–1914*. Oxford: Basil Blackwell, 1981.

Swett, Pamela. *Neighbors and Enemies: The Culture of Radicalism in Berlin, 1929–1933*. Cambridge: Cambridge University Press, 2004.

Taylor, Charles. *A Secular Age*. Cambridge, MA: Belknap Press, 2007.

Thienel, Ingrid. *Städtewachstum im Industrialisierungsprozess des 19. Jahrhunderts: Das Berliner Beispiel*. Publikationen zur Geschichte der Industrialisierung 3. Veröffentlichungen der Historischen Kommission zu Berlin 39. Berlin: Walter de Gruyter, 1973.

Thomason, Frank J. "The Prussian Police State in Berlin, 1848–1871." PhD dissertation, Johns Hopkins University, Baltimore, Maryland, 1978.

Thum, Gregor. "Imperialists in Panic: The Evocation of Empire at Germany's Eastern Frontier around 1900." *Helpless Imperialists: Imperial Failure, Fear, and Radicalization*, edited by Maurus Reinkowski and Gregor Thum, 137–62. Göttingen: Vandenhoeck and Ruprecht, 2013.

Tippach, Thomas. *Koblenz als preussische Garnison- und Festungsstadt: Wirtschaft, Infrastruktur und Städtebau*. Cologne: Böhlau Verlag, 2000.

Topfstedt, Thomas, Hartmut Zwahr, and Katrin Löffler, eds. *Leipzig um 1800: Beiträge zur Sozial- und Kulturgeschichte*. Beucha: Sax-Verlag, 1998.

Tracy, James D. "To Wall or Not to Wall: Evidence from Medieval Germany." In *City Walls: The Urban Enceinte in Global Perspective*, edited by James D. Tracy, 71–87. Cambridge: Cambridge University Press, 2000.

Tuan, Yi-Fu. *Space and Place: The Perspective of Experience*. Minneapolis: University of Minnesota Press, 1977.

Uhlenhuth, Theodor. "Die Paderborner Stadtmauern und ihr Abbruch." *Heimatborn* 11, nos. 2, 3, and 4 (1931): 5–6, 9–10, 13–14.

Umbach, Maiken. *German Cities and Bourgeois Modernism, 1890–1924*. Oxford: Oxford University Press, 2009.

Umbach, Maiken. "A Tale of Second Cities: Autonomy, Culture, and the Law in Hamburg and Barcelona in the Late Nineteenth Century." *American Historical Review* 10, no. 3 (June 1, 2005): 659–92.

Urban, Florian. "The Hut on the Garden Plot: Informal Architecture in Twentieth-Century Berlin." *Journal of the Society of Architectural Historians* 72, no. 2 (June 2013): 221–49.

Vasudevan, Alexander. *Metropolitan Preoccupations: The Spatial Politics of Squatting in Berlin*. Chichester, UK: Wiley Blackwell, 2015.

Voyles, Traci Brynne. *Wastelanding: Legacies of Uranium Mining in Navajo Country*. Minneapolis: University of Minnesota Press, 2015.

Wade, Richard C. *The Urban Frontier: The Rise of Western Cities, 1790–1830*. Cambridge, MA: Harvard University Press, 1959.

Walker, Mack. *German Home Towns: Community, State, and General Estate, 1648–1871*. Ithaca, NY: Cornell University Press, 1971.

Walker, Mack. *Germany and the Emigration, 1816–1885*. Cambridge, MA: Harvard University Press, 1964.

Ward, Colin. *Autonomy, Solidarity, Possibility: The Colin Ward Reader*. Edited by Chris Wilbert and Damian F. White. Oakland, CA: AK Press, 2011.

Waterhouse, Alan. *Boundaries of the City: The Architecture of Western Urbanism*. Toronto: University of Toronto Press, 1993.

Weber, Karl-Klaus. "Stadt und Befestigung: Zur Frage der räumlichen Wachstumsbeschränkung durch bastionäre Befestigungen im 17. und 18. Jahrhundert." *Die alte Stadt* 4 (1995): 301–21.

Wegmann-Fetsch, Monika. *Die Revolution von 1848 im Grossherzogtum Oldenburg*. Oldenburger Studien 10. Oldenburg: Heinz Holzberg Verlag, 1974.

Wein, Martin. *Stadt wider Willen: Kommunale Entwicklung in Wilhelmshaven/Rüstringen 1853–1937*. Marburg: Tectum Verlag, 2006.

Wilson, Jeffrey K. *The German Forest: Nature, Identity, and the Contestation of a National Symbol, 1871–1914*. Toronto: University of Toronto Press, 2012.

Wood, Nathaniel D. *Becoming Metropolitan: Urban Selfhood and the Making of Modern Cracow*. DeKalb: Northern Illinois University Press, 2010.

Zahra, Tara. *The Great Departure: Mass Migration from Eastern Europe and the Making of the Free World*. New York: W. W. Norton, 2017.

Zelljadt, Katja. "Presenting and Consuming the Past: Old Berlin at the Industrial Exhibition of 1896." *Journal of Urban History* 31, no. 3 (March 1, 2005): 306–33.

Zimmer, Oliver. *Remaking the Rhythms of Life: German Communities in the Age of the Nation-State*. Oxford: Oxford University Press, 2013.

Zschocke, Helmut. *Die Berliner Akzisemauer: Die vorletzte Mauer der Stadt*. Berlin: Berlin Story Verlag, 2007.

Zwahr, Hartmut, Thomas Topfstedt, and Günter Bentele, eds. *Leipzigs Messen 1497–1997, Teilband 1: 1497–1914*. Geschichte und Politik in Sachsen. Cologne: Böhlau Verlag, 1999.

Index

Note: Page numbers in *italics* refer to figures.